Bioburst

Bioburst

The Impact of Modern Biology
on the Affairs of Man

RICHARD NOEL RE

Louisiana State University Press

BATON ROUGE AND LONDON

Designer: Christopher Wilcox
Typeface: Times Roman
Typesetter: G & S Typesetters, Inc.
Printer: Thomson Shore, Inc.
Binder: John Dekker & Sons, Inc.

The passage quoted herein from Martin Heidegger's *Discourse on Thinking*,
translated by John M. Anderson and E. Hans Freund (copyright © 1966 by
Harper and Row, Publishers, Inc.), is reprinted with the gracious permission of
Harper and Row, Publishers, Inc.

Library of Congress Cataloging-in-Publication Data

Re, Richard Noel.
    Bioburst: the impact of modern biology on the
affairs of man.

    Bibliography: p.
    Includes index.
    1. Biology—Social aspects.   2. Biology—
Philosophy.   3. Molecular biology—Social aspects.
4. Molecular biology—Philosophy.   I. Title.
QH333.R4   1986   303.4'83   86-7422
ISBN 0-8071-1289-5

To my family, my teachers,
and all others who love inquiry

Contents

Figures

Figures

A Note to the Reader

These essays deal with the life of the body and the life of the mind. They contain fact and speculation, each clearly identified. Because technical material has been kept to a minimum, the interested reader is referred to the notes where subthemes are explored. Finally, it must be recalled that molecular biology has proven itself to be an extremely rapidly developing scientific discipline, and it is to be expected that even in the near term many of the technical aspects of this work will be expanded by, or supplanted by, new knowledge. I hope, however, that the major themes of these essays will serve the reader well over time.

Preface

In the warm humid spring of 1721 HMS *Seahorse,* newly arrived from the West Indies, dropped anchor in Boston harbor. After the ship spent several weeks in port, it became apparent that smallpox had broken out on board. Not surprisingly, guards were dispatched to quarantine the vessel, which was soon towed to the outer harbor for the purpose of improving isolation. But all precautions notwithstanding, cases of smallpox appeared among the townspeople and these were immediately quarantined. Soon the number of cases grew so rapidly that guards were removed from the sick houses and the populace accepted the inevitability of epidemic.

For all the fear and despair rampant among the citizenry, the epidemic was not totally unexpected, for the disease occurred with some regularity in Massachusetts. Indeed, there were those who had intellectually prepared to combat the event. Almost as soon as the epidemic was recognized, the Reverend Cotton Mather, a Boston preacher of some repute, joined forces with his friend and medical colleague, Dr. Zabdiel Boylston, to commend the practice of variolation to the townspeople. Indeed, Mather, having borrowed a tattered copy of the *Transactions of the Royal Society of Medicine* from one of Boston's more conservative physicians, had learned enough of the technique to permit Boylston to variolate his son and a manservant in the very early days of the epidemic.

Variolation, the deliberate inoculation of healthy patients with pus from smallpox sores, had been employed intermittently and admittedly infrequently by the Turks, Africans, and some Caribbean tribes. A wick dipped in smallpox pus was usually inserted in a small puncture wound in the arm and left in place for twenty-four hours. Thereafter, the patient developed a mild illness associated with fever and a few smallpox sores localized to the arm. From this he usually recovered rapidly to go forth forever immune to smallpox. Sometimes he died promptly. Indeed, when earlier in 1721 smallpox had appeared in England, variolation was attempted, but the two deaths in variolated subjects that greeted the inception of the project quickly dampened popular enthusiasm for the practice.

This was not the case in Boston. Although many townspeople objected strenuously to the idea—an attempt was made to bomb Mather's home, and the selectmen of the city sought to indict Boylston—variolation was continued by virtue of the perseverance of Boylston and Mather. After several months, the public mood began to change, in large part because of the statistical evidence collected by Boylston. Although 6 of Dr. Boylston's 286 inoculees died for a case fatality rate of about 2 percent, the fatality rate from smallpox itself was substantially higher at 15 percent. The utility of the procedure was proven, and public approval soon followed. When after some months the epidemic ended, there were those who wished to continue the practice prophylactically. However, cooler heads prevailed and variolation was stopped.

In the epidemic's aftermath, Boston's learned citizens began to debate not so much the efficacy of the procedure as its morality. Was it right to make some who might never have contracted smallpox fatally ill with variolation? Was it right for anyone deliberately to make himself ill for any reason? Were there long-term ill effects associated with the procedure? Should variolated patients be quarantined to protect the unvariolated? In various writings describing the 1721 epidemic, Cotton Mather discussed these issues at length and with surprising clarity. When it was all over, the people of Boston came to realize that they had gained a certain measure of control over smallpox, and with that power came new insights, not only into the mechanism of disease, but also into their conception of life and morality. They had learned of the potential and the dangers of biological intervention. They were in a very real sense richer by virtue of the lessons of the epidemic. No matter that not long after 1721 the use of the substantially (but not completely) safer cowpox virus (vaccination) for protection against smallpox appeared to render moot the lessons of variolation, the lessons were learned and learned well.

No one in 1721 could have anticipated that these lessons would find poignant modern parallels in the contamination of the early polio vaccines with the animal-cancer-causing virus SV-40, in the swine flu vaccination debacle of 1976, and perhaps in the AIDS epidemic of the present. No Boston colonial could have known that the scientific, moral, and societal issues that were raised—and answered—by the variolation question of 1721 would be raised again today, albeit in a more technical garb, and applied to genetic engineering and recombinant DNA technology. And, for our part, it remains to be seen if the lessons of variolation will be fully recalled and applied in the context of the present day.

The variolation experience taught not only the eighteenth-century medical community, but laymen as well, a great deal about the world in which they lived, and in many ways it dramatically changed their view toward that world. And yet compared to the potential impact of current advances in molecular biology, the lessons of variolation pale, for perhaps at no other time in the history of medicine has the potential existed for bioscience to influence the world view of man so radically as it does today. Specifically, recent advances in the understanding of nucleic acid chemistry, gene regulation, hormone action, and immunology are not only leading to a technological explosion that is daily reported by the news media, but more important though less obviously, they are producing a deeper appreciation of the world in which we live. To the ramifications of these insights, I affix the term *Bioburst*, implying that a rapidly expanding horizon of biological and philosophical understanding will surely follow from them. The mosaic produced by these accumulating insights and the relentless truth that supports them will surely have a major impact not only on medicine, agriculture, and commerce, but equally upon society's modes of thought. Bioburst will change in manifestly important ways the way we think about our world and ourselves, about our children and our parents, about the living, the nonliving, and the immortal. It is the goal of these essays to explore the ramifications of Bioburst.

Special thanks are due to Juanita C. Shipman and Carolyn F. Davis for typing the manuscript, and to Marion Stafford and Margaret Fisher Dalrymple for their diligence in editing it. Barbara Siede was most helpful in preparing the illustrations that accompany the text. Finally, I would like to acknowledge, as well as express my appreciation for, the help and encouragement given me by the entire staff of the Louisiana State University Press.

Bioburst

CHAPTER I

Cotton Mather: A Molecular Biologist in Colonial America

Cotton Mather, the so-called "Puritan priest," is a man familiar to all historians interested in early America. Undeniably, he had a profound impact on the intellectual climate in America during the late seventeenth and early eighteenth centuries. In placing this man in context, it is important to realize that the era into which Mather was born in 1663 was what might be called medieval early America, a time very different from the post-Enlightenment America characterized by Georgian architecture and republican ideas. Nonetheless, it was a time of considerable philosophical evolution, and Cotton Mather was a prominent force not only in the religious and legal thought of the day, but also in medical affairs.

Cotton Mather's maternal grandfather was John Cotton, who for twenty years was a minister of St. Botolph Church in Boston, Linconshire, England. He was compelled to give up his post when a townsman charged that magistrates were not required to kneel at the sacrament in Reverend Cotton's church. In some theological disgrace, John Cotton set off for the New World, and, as it happened, he sailed for America with two other ministers, Thomas Hooker and Samuel Stone. This was taken to be a good omen by all three, for one ship brought Cotton for clothing, Stone for building, and Hooker for fishing. Reverend Cotton preached in Boston for nineteen years until his death in 1652. His widow married Richard Mather.

Mather was also a minister who had been suspended for nonconformity and came to New England as a result. His first wife was

Catherine Holt, of whom Cotton Mather was later to write: "if a pretty late abortion might have passed for a birth, it might be said of this gentle woman that she was a mother of seven sons."[1] The last of these sons was Increase Mather. He too entered the ministry and began preaching in the Second Church of Boston. On March 6, 1662, he married the daughter of John Cotton. To this couple Cotton Mather was born on February 12, 1663. Increase Mather later became one of the most prominent early Bostonians and was named president of Harvard College in 1685.

Cotton Mather wished to enter the clergy like his father, but because he was troubled with stuttering he elected to abandon the ministry, at least temporarily, and take up medicine. He was admitted to Harvard College and applied for a bachelor's degree in 1678—at fifteen years of age the youngest graduate to that date. Eventually he married Abigale Phillips, who bore him nine children, five of whom died in childhood. He mastered his stuttering and eventually became assistant minister at his father's church.[2]

The Mathers were at the height of their prestige when, in 1692, witchcraft erupted in Salem. The frenzy began when the children of Mr. Paris, the minister of Salem village, were seized early in the year with disorders of "no earthly origin." A special court consisting of six judges under Deputy Governor William Stoughton was established to try the witches allegedly responsible for the ephemeral epidemic. On June 2, 1692, Bridgett Bishop was condemned, and on June 10 she was executed for witchcraft. Before the townspeople of Salem went further, they consulted the ministers of Boston, who issued a reply that addressed the importance of circumspection in the methods of examination employed against witches but essentially recommended that the proceedings continue. Nineteen people were eventually hanged, one was pressed to death for failure to answer his indictment (the only such case in American jurisprudence), two died in jail, and hundreds were arrested. Eventually, on August 19, 1692, the Reverend George Burroughs, a Harvard graduate, was hanged. On the gallows, Burroughs eloquently protested his innocence with great effect upon the crowd. Cotton Mather was present and launched into a diatribe against the minister, winning the day, and bringing about Burroughs' hanging. On

September 22, the last witch was hanged, and the trials thereafter halted because the accusations were, by weight of numbers, losing all credibility. Indeed, the next year a woman in Boston became "possessed" but was promptly ridiculed.[3]

Cotton and Increase Mather defended their role in the Salem affair. Cotton wrote that "flashy people may burlesk these things, but when hundreds of the sober people in a country where they have as much mother wit certainly as the rest of mankind, know them to be true, nothing but the absurd and the froward spirit of Sadducism can question them."[4] This defense notwithstanding, the ridicule and disapproval that descended upon the Mathers were severe. And then in 1702, as the result of a complex interplay of political and religious forces, Increase was compelled by his enemies to resign as president of Harvard College. Cotton's prestige transiently waned, and he turned more and more toward natural science. In 1712, he was invited to send letters to the Royal Society of London as a correspondent. This he did in great abundance, and in 1713 he was elected a fellow of the society. In that year an epidemic of measles occurred in Boston, and Mather rose to the occasion by publishing a pamphlet entitled *A Letter, About a Good Management Under the Distemper of the Measles at This Time Spreading in the Country, Here Published for the Benefit of the Poor and Sick As May Want for the Help of Able Physicians.* This document was not well received by the physicians of the day, who perhaps feared competition from what amounted to a book of home remedies.

In 1716, Cotton Mather, having read Manuel Timoni's account of the Turkish practice of variolation, came upon the account of Jacob Pylarini reporting similar successes. Mather claims to have had some independent knowledge of this procedure, saying that a West Indian manservant had informed him of the usefulness of the practice some time before. He resolved to try it should smallpox again strike Boston. In a letter of July 12, 1716, Mather wrote, "for my part, if I should live to see the Small-Pox again enter into our City, I would immediately procure a Consult of our Physicians, to Introduce a Practice, which may be of so very happy a Tendency."[5]

In 1721, smallpox appeared in London. At the instigation of one Lady Montegu, twenty people were inoculated by the technique of variolation and two died. Thereupon the practice at least temporarily fell into disfavor. An infected ship arrived in Boston early in the spring of 1721. During May and June, the vessel was kept down harbor and guards were placed around all sick houses. Streets were cleaned and sealed. By mid-June the illness reached epidemic proportions and the guards were removed. About this time, Mather wrote a letter to the physicians of Boston informing them of the reports of Timoni and Pylarini and suggesting that they begin to experiment with variolation. This proposal was rejected out of hand by the medical community. Only Dr. Zabdiel Boylston saw merit in the idea.[6]

On June 26, Boylston inoculated his son and two Negro slaves. He wrote that the procedure was carried out by making several incisions in the arm into which bits of lint dipped in pox matter were inserted. After twenty-four hours the lint was withdrawn and the wound dressed with warm cabbage leaves. On the seventh day the sickness began and pustules appeared, sometimes few, sometimes hundreds. The pox matter itself was collected from sores eleven, twelve, thirteen, and fourteen days after the onset of the disease and stored cool until use. Boylston wrote, "so long as it [the pox matter] retains its white color and even consistence and that without smell, you may depend that it is good."[7]

A local physician, Dr. William Douglas—the only man in Boston with a medical degree from a European institution—attacked Boylston as a quack, noting that Boylston had no medical degree at all. Ironically, it was Douglas' copy of the *Transactions of the Royal Society* that Mather had borrowed in order to read the works of Timoni and Pylarini, a fact that in no way endeared either Mather or Boylston to him. Douglas adamantly objected to the practice of variolation and on July 12 attacked it in the Boston *Newsletter,* citing three objections. First, he deemed it dangerous, which was not an altogether untenable position. In the second place, he felt it irreligious to interfere with God's providence in these matters. While this might be seen as an attempt to marshal ministerial support on his side, it must nonetheless be viewed as a remarkable position for a physician to take. Finally, Douglas argued that Boylston's use of variolation was criminal from the point of view of the public, representing as it did a private undertaking with major implications for the public health. However, six ministers, among them Increase Mather, Cotton Mather, and Benjamin Colman, replied in the *Gazette* of July 31, 1721, defending Boylston and deeming the practice moral. Soon the New England *Currant,* a paper run by James and Benjamin Franklin and later noted for its anticlerical views, took the side of the physicians against the ministers and Boylston.[8]

Feelings ran high. In November a bomb was thrown into Cotton Mather's house. Fortunately, the bomb did not explode and no one was injured. Mather described the event as follows: "the mercyful providence of my Savior so ordered it, that the Grenado passing thro' the window had by the iron in the middle of the casement, such a turn given to it, that in falling on the floor, the fired wild-fire in the fuse was violently shaken out upon the floor, without firing the Grenado. When the *grenado* was taken up, there was found a paper so tied with string about the fuse that it might out-live the breaking of the shell—which had these words in it—Cotton Mather, you Dog; Dam you: I'l enoculate you with this, with a pox to you."[9]

By December, 250 people had been inoculated. Not long thereafter, the epidemic ended. A total of 280 people were variolated, 274 were clinically felt to have developed attenuated smallpox by virtue of variolation, and 6 died of the inoculation. Boylston reasonably argued that many of those who died subsequent to variolation may have been previously infected by a natural route prior to inoculation with smallpox pus. This possibility could not be excluded because Boylston and his co-investigators were unwilling to quarantine patients long enough to be certain that exposure had not occurred prior to vaccination. Although isolation prior to variolation would have made for a better experiment, it was not undertaken for reasons of morality as well as practicality. The experimenters knew that deaths which were in fact not caused by variolation would be ascribed to their procedure, but they chose to assume this liability rather than deprive patients of potentially lifesaving treatment. These investigators were aware of the scientific issues involved and opted to act on the basis of what they perceived to be the moral course. Why Boylston elected not to isolate variolated patients so as to protect the unvariolated from this new potential source of infection is less clear. Arguably, the additional risk associated with the presence of 200 or so variolated people was negligible during an epidemic. Perhaps issues related to practicality were uppermost in the minds of Boylston and Mather. In any event, the mortality after inoculation was 2.5 percent or less, compared to 15 percent in naturally acquired cases. Although some authors disputed the early results, Boylston's account was eventually accepted and stands as probably the best medical description of the results of variolation. In the spring of 1722, he was ordered to stop the procedure because the epidemic had abated.[10]

However, many issues were left unresolved, and charges continued to fly back and forth even after the passing of the smallpox. Cotton Mather on numerous occasions answered the critics of variolation, rebutting in the process several important arguments. The first was the possibility that future harm would be done to the patients undergoing variolation by unanticipated side effects of the procedure. Mather's reply was "a crazy old man that is near 70 having lately enjoyed the benefits of inoculation 'tis thought that if he should die one minute before 90, these people (if not come to their senses before) will say this inoculation kill'd him." To a second point raised by the critics of variolation that "the whole have no need of a physician and that it is not lawful for me to make myself sick when I am well and bring the sickness on myself though it be to prevent a greater sickness," Mather replied that " 'tis no purpose to tell them that they cavil against the use of all . . . physic and that they confute themselves as often as they take vomit or use a blister." A third point raised against variolation was that it was not reasonable to try a thing learned from heathens, and Mather quickly replied by asking from whom his critics had learned the teaching of Hippocrates and Galen, and from whom they had learned to smoke tobacco, or drink tea and coffee.[11]

Again and again Mather and his associates returned to the moral questions raised by variolation in works like *A Reply to the Objections Made Against It from Principles of Conscience*. On the title page of this monograph was written, "then Jesus said unto them, I will ask you one thing, is it lawful to save a life or destroy it?" The most serious moral objection to variolation, of course, was the problem of spreading the contagion by deliberately exposing people to the disease. It had been immediately appreciated by Boylston and Mather that inoculation could lead to a wider epidemic with increased mortality among unvariolated citizens, yet they had not isolated their subjects. This public health question was debated at great length, but in the end the community resolved that variolation was a worthwhile practice—however repulsive it might seem esthetically and morally. Indeed, the prevalent opinion was that, should smallpox recur, the practice would be reinstituted. However, it was also more or less the consensus that, should future variolation be undertaken, variolated patients would be isolated for the protection of the community. Here was the beginning of an appreciation of the idea that the public good could place constraints on medical experimentation, particularly when those undertakings involved exposing the public to clearly contagious agents. The failure to isolate variolated subjects was the major oversight of the entire Boylston-Mather experiment and without doubt it posed the greatest danger to the public.[12]

Variolation proved quite successful and eventually became fashionable, remaining so even after Jenner introduced vaccination, the procedure that until recently has been widely employed for the prevention of smallpox. For his part, Cotton Mather went on to publish other great works dealing with his views on medicine, including *The Angel of Bethesda*. However, his greatest medical achievement was the role he played in the introduction of variolation to Boston. As one of Mather's biographers has pointed out, "the history of immunology, with all its ultimate values in overcoming infectious diseases, began—above the folk level and on a meaningful scale—in the Boston of 1721." Mather's speculations on the mechanisms of infectious diseases and immunization are also of importance and display a modern attitude toward experimentation:

> Were one of an ordinary capacity . . . willing to try a little how far *Philosophy* might countenance the Matter: One might think, the venemous *Miasms* of the *Small-Pox* entering into the Body, in the way of *inspiration*, are immediately taken into the Blood of the *Lungs;* and, I pray, how many Pulses pass before the very *Heart* is pierc'd with them?
> And within how many more they are convey'd into all the *Bowels*, is

easily apprehended, by all who no any thing how *Circulation* of the *Blood* is carry'd on; at the same time the *Bowels* themselves are infeebled, and their tone impair'd, by the *Venom* that is thus insinuated. Behold the Enemy at once got into the very *Center* of the Citadel; and the invaded Party must be very strong indeed, if it can struggle with him, and after all entirely expel and conquer him: Whereas the Miasms of the *Small-Pox,* being admitted in the Way of *Innoculation,* their Approaches are made only by way of the *Out Works* of the Citadel and at a considerable Distance from it. The Enemy, 'tis true, gets in so far, as to make some *Spoil;* even so much as to satisfy him, and leave no *prey* in the *Body* of the Patient, for him ever afterwards to seize upon; but the *Vital Powers* are kept so clean from his assaults, that they can manage the *Combat* bravely; and tho' not without *surrender* of those Humours in the Blood, which the Invader makes a Seizure on, they oblige him to March Out the same Way *he came in,* and are sure of never being troubled with him any more. If the *Vermicular* Hypothesis of the *Small-Pox* be receiv'd with us, (and it may be, as many now think, an *animaculated* business) there is less of *Metaphor* in our Account, than may be at first imagin'd.

But to what purpose is all this *Jargon?* And of what significancy are most of our *speculations?* EXPERIENCE! EXPERIENCE! 'tis to thee that the Matter must be referr'd after all; a few *Empericks* here, are worth all our Dogmatists.[13]

Mather's animalcular theory of smallpox was lost on subsequent generations—until the discovery of the virus and an understanding of its function revealed the true nature of the disease. Mather and Boylston not only performed the first major public health undertaking in the New World and developed mechanisms for analyzing the results of the enterprise, they additionally added to the store of basic scientific knowledge in immunology and infectious diseases. Mather and his colleagues also argued, albeit from a religious point of view, for all those scientists who would follow them that there can be no benefit without uncertainty. Their admonition was "if you come into the Practice [of variolation], I know you will not do it in *Carnal Security;* for that may provoke God to deny the blessing." In this, the variolators of eighteenth-century Boston foreshadowed Gödel, Heisenberg, Shannon, and others who pointed out the indeterminant nature of the world in which we live.[14]

Even the most cursory analysis of the life of Cotton Mather cannot fail to reveal the starkly contrasting roles he played in the Salem witch trials and in the variolation process of 1721. Although in each case he fiercely followed his beliefs, one gets the distinct impression that he learned a great deal between the two episodes. By implication, he must have learned much about the nature of reliable evidence, about the nature of illness, and about man's interaction with both evidence and illness. He understood that morality did not require passivity in the face of biological danger, and he used the same courage he had summoned to condemn Reverend Burroughs in a newer, more productive way to limit smallpox in Boston. With this courage he defended the morality of science and medicine while eventually, along with Boylston, coming to believe that public oversight is a critically important facet of any widespread biological intervention.

The works of Boylston and Mather cry out for study lest we be deluded in the belief that the moral and public-health questions swarming about modern biology are newly raised by the technology of genetic engineering. Certainly, recombinant DNA technology does raise complex issues, both moral and political, but these issues are really elaborations or variations on ideas and fears raised many times in the past. In a real sense, Mather and Boylston were genetic engineers of a sort much more intrepid than any scientist living today. They were embroiled in a contest between a clergy more powerful and a medical profession more conservative than the religious and medical establishments of today. And in the interaction of these two camps—a contest that took place more than two hundred fifty years ago—are to be found paradigms that will provide the answers to many of today's questions and problems.

CHAPTER II

Ground Rules, Terms, and Codes

Nothing is perhaps so hard to define as life, unless it is consciousness, with which we shall contend in a later chapter. Traditionally, a physical-chemical entity is defined as alive if it is capable of self-replication and metabolism. By the latter term is meant the conversion of chemicals in the environment into new compounds or into energy. For example, yeast are capable of self-replication as well as of a host of metabolic functions, not the least of which is the conversion of sugar into alcohol. This chemical transition is, of course, of great value to the viniculture industry since it lies at the very heart of wine production. But additionally, it is of no mean importance to yeast, which derive a considerable amount of energy from the process. The energy liberated can be used for a variety of cell functions, such as growth and replication. Clearly, the yeast cell is a living entity since it has the capacity for self-replication and metabolism. It should be noted, however, that even the classical definition of life does not require that an entity be self-replicating or metabolizing at the time of observation to be deemed alive. The potential is enough. Prepubertal children do not replicate but nonetheless are alive. Spores are primitive organisms that lie dormant—without metabolism—for long periods. So do seeds. If only the potential for metabolism is required, then both seeds and spores can be considered alive.

The issue becomes trickier when one considers viruses. In essence, viruses are crystalline protein surrounding a bit of the genetic material DNA or RNA. The protein coat of the virus has the capacity to bind and fuse with the outer membranes of living cells, thereby producing a microscopic hole in the membrane through which the genetically active DNA or RNA is injected into the cell. Once inside, the genetic material of the virus either lies relatively dormant for variable and often long periods of time, reproducing itself in coordination with the normal division of the cells so that each progeny of the infected cell is itself infected, and only slowly making additional virus to spread the infection—or the viral genetic material may completely take over the cell, subvert the normal cell machinery totally to the task of making immense quantities of virus, kill the cell, and finally unleash a new wave of virus upon neighboring cells. Viruses can thus kill or disable, and this is one reason why we are afraid of them (an even more intimidating feature of viruses will be discussed later).[1] One can legitimately question whether a virus is alive, for although it can replicate and produce metabolic change, it can do so only with the assistance of a living (by any definition) entity—namely the cell it infects.

This question of the status (living or nonliving) of what essentially amounts to an intracellular chemical parasite is even more poignant in the case of the recently discovered viroids. These entities are long RNA molecules bent over upon themselves in a clothespinlike configuration (a shape that renders them extraordinarily resistant to disinfectants, boiling, and even x-ray). Viroids commonly infect plants, and produce disease and often death as a result. Recently, it has been demonstrated that *kuru,* a severe degenerative disease of the central nervous system found in certain New Guinean communities that ritualistically eat the brains of deceased family members, is related to infection with some "primitive" entity—either a primitive virus, an atypical viroid, or, as some believe, a so-called "proteinaceous infectious particle," or *prion.* The prion hypothesis arises from the fact that no one yet has been able to isolate either RNA or DNA from the primitive infectious agent that causes *kuru.* Although many scientists doubt that these prions exist, the possibility is intriguing. How a prion could reproduce is a matter of some speculation and potentially of great importance. One theoretically possible way a prion could replicate is by entering a cell and causing the cell's own gene coding for the specific prion protein to

11

make it in excess. This process is akin to bringing a microphone near an amplifier and producing a loud, nerve-wracking oscillation. According to our hypothesis, however, the oscillations in protein synthesis produced by prions would literally be nerve-wracking, leading to the destruction of the patient's central nervous system. In any case, it is clear that *kuru* is caused by some fairly primitive species or "chemical." This infectious particle apparently survives in preparations of brain that are eaten by victims of the disease. The particle spreads to the brain of its new human host and slowly, systematically, begins destroying brain function. A paralyzed, seizure-wracked, demented patient is the pathetic result.

Recently, it has been appreciated that a particularly severe form of senility in man (Creutzfeldt-Jakob disease) likely results from chronic infection with a similar primitive agent. Transmission from one human to another has been reported under unusual circumstances. Specifically, when the cornea of a patient who died from Creutzfeldt-Jakob disease was transplanted in a sight-saving operation into a healthy patient, the recipient was found some time later to have developed the disease also. The infectious entity was subsequently isolated from human material and found to be extraordinarily resistant to usual disinfectant procedures—an observation that caused considerable chagrin in the neurosurgical community, since the theoretical possibility of transmission of the disease from one person to another via neurosurgical instruments seemed frighteningly real. Potent new procedures for disinfection of surgical instruments were the result. Even more recently, evidence has surfaced to suggest that viruses or prions could be involved in processes leading to Alzheimer's disease—another tragic form of premature senility.[2]

Let us return to the central question: should viruses that have a potentially enormous impact on human health and disease, and viroidlike entities and proteinaceous infectious particles (if they actually exist) that apparently can incapacitate plants and even man, albeit less frequently, be considered alive or dead? It would seem reasonable, by virtue of their impact on living systems and by virtue of the fact that life as we know it may have evolved from bits of chemically self-replicating long molecules like viroids, that these viruses and viroids should be considered alive. Certainly, *if* all life could be shown to have sprung from them, they necessarily must have (given time and the right conditions) the potential for self-replication and metabolism. And it is possible that forms such as these did indeed give rise to life.

Therefore, for the purposes of this discussion, living things will be divided into two classes (taking considerable liberties with current classifications). The first class, "the major forms of life," will be defined as consisting of entities that are capable of self-replication and metabolism if placed in a suitable chemical environment—but which do not require for these functions the component parts, the machinery as it were, of any other living species. The second class, "the minor forms," are those that can multiply and metabolize only with the help of other species of life. The distinction between the two classes would at first glance appear self-evident. A man is a major form, whereas a viroid is a member of the minor class. Of course, the distinction becomes a bit murky when one considers that man requires the oxygen and starch produced by plants in order to live. Were there no plants, the planet would soon have an atmosphere quite inhospitable to man; were there no plant-mediated photosynthesis, there would be no food of any sort for man, and the abundant energy of sunlight would no longer be harnessed to the support of life. Thus, man is dependent on plants for his self-replication and metabolism, but he doesn't have to use the chemical machinery of the plant itself. He lives on a product of the plant and therefore is still classified as a major organism by the system adopted here.

The issue becomes stickier in the case of the mitochondrion. As best science can tell, mitochondria, the ovoid entities that generate the energy of animal cells, were once free-living organisms that took up residence inside living cells, forming something of a partnership. They provide metabolically derived energy (more formally, they generate the high-energy compound adenosine-triphosphate or ATP) to the cell, while the cell provides them with a hospitable chemical environment. At first glance, this would appear to be a symbiotic relationship—that is, a case of parasitism that helps both parties. The mitochondria multiply independently of the whole cell and seem therefore to fill both criteria for classification as living organisms, since as far as chemical generation of energy is con-

13

cerned, they are certainly the metabolic stars of the living world. Yet do they really meet the definition? It turns out that, while mitochondria can divide on their own, some of their genes are stored in the nucleus of the host cell. Mitochondria do not contain their entire gene pool but to some extent share the genes of the host cell. The question is therefore left to the reader as to whether mitochondria should be considered living entities or living parts of cells. To the sophisticate, the question may sound trivial, and yet a little reflection upon the status of the mitochondria may influence one's attitude toward parasitism, infection, and even cancer, as well as toward patients who suffer from these conditions, for in essence we may all be parasitized in a very fundamental way.[3]

The epistemological purist, of course, could argue that the differentiation of living from nonliving entities is moot, since life apparently evolved from nonlife and therefore any distinction between the two is artificial. This argument carries considerable intellectual force, but for our purposes it will not be totally adopted because any discussion of the impact of biology upon the affairs of men requires that some attention be paid to the traditional concepts of life. After all, society must take a different view of the deliberate destruction of an adult member of the species *Homo sapiens* (a man) than it would take of the destruction of a large stone. Indeed, it must, and does, view differently the destruction of a laboratory animal such as a rat and the destruction of a rock. The purist argues that the rat is a more intricate and more valued (by man) bit of the universe than is the rock and so is of more concern to us. Life in this sense becomes in part a question of value to the mind defining it, and in part a question of chemistry; and the virus becomes the interface between the living and the nonliving.

There is a lot in this view with which the traditionalist can agree. The traditionalist perhaps places more emphasis on the pluripotentiality of DNA than on other nonreproducing chemical polymers and therefore is prone to classify the virus as being alive. But the purist's and traditionalist's views are not so widely divergent, since both hold that the virus represents the interface between two very different forms of matter. One can speculate that the purist, if pressed, would admit that the existence or nonexistence of self-replicating bits of DNA (life) neither adds nor detracts a great deal from the enormous complexity and wonder of the cosmos. Man is one element of the cosmos, perhaps equivalent in complexity to a spiral galaxy. It is then for largely egotistical reasons that the more traditional view is adopted here. We should note, however, that at least one school of thought holds that living, conscious matter is physically distinct from inanimate matter in the way it interacts with the cosmos.[4]

Having provided an imperfect though reasonably satisfactory definition of life, several potentially interesting questions and observations can be generated. First, there appears to be a remarkable economy of mechanism by which the variety of living forms commonly encountered on this planet operate. The question of how forms on other planets operate will be deferred for the moment, but in fact this is a fertile field for theorization. Common living forms on earth make use of virtually all normally occurring elements in the earth's environment but employ predominantly carbon, hydrogen, oxygen, nitrogen, and phosphorus in their major components. These elements are functionally grouped into a myriad of different classes of molecules, but predominant among them are proteins, carbohydrates, lipids, steroids, and nucleic acids. In general, proteins supply structural support for organisms, regulate chemical reactions in living cells, and finally can be used as an energy source. Carbohydrates are a major energy source for living things and also provide structural support. Lipids and steroids provide structural elements for cells, have some metabolic control functions, and lipids can also be used as energy sources. Finally, nucleic acids provide the blueprint for making new organisms (replication) as well as for the regulation of cell functions (metabolism). It is the relationship of nucleic acids and proteins, the correlation or correspondence of nucleic acid sequences with protein sequences, that lies at the heart of all life on earth.[5]

The so-called bases that are the distinctive components of nucleic acids occur in nature in five basic forms—adenine, thymine,

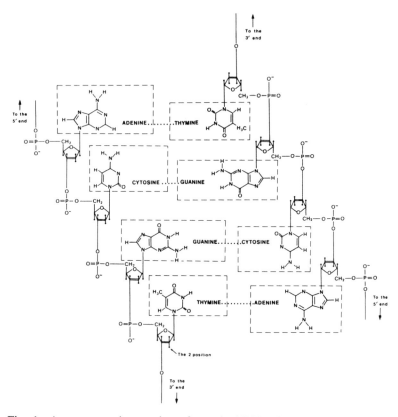

Fig. 1. A representative portion of a typical DNA double helix. The
molecule consists of two antiparallel chains (*i.e.*, one running 5′ to 3′,
the other 3′ to 5′) held together by the binding of thymine with adenine
and guanine with cytosine. Each DNA chain consists of bases, shown in
the boxes, joined to a deoxyribose-phosphate backbone. The 2 position
of deoxyribose is indicated by an arrow. If the hydrogen atom found at
this position is replaced on each sugar by OH (hydroxyl) groups, the
chain becomes a ribonucleic acid (RNA) instead of a *deoxy*ribonucleic
acid (DNA). RNA is usually found as a single chain molecule and contains the base uracil instead of thymine.

guanine, cytosine, and uracil. These are small, complex molecules formed of carbon, hydrogen, nitrogen, and oxygen. In living organisms, these bases are usually found chemically bound to pentagonally shaped sugars to form nucleosides or deoxynucleosides, depending upon whether an oxygen molecule is found at the 2 position of the sugar (in which case we are dealing with a nucleoside) or is absent (in which case we are dealing with a deoxynucleoside) (see figure 1). When a nucleoside or deoxynucleoside is bound to a phosphate group, the entire structure consisting of base, sugar, and phosphate is termed a *nucleotide* or *deoxynucleotide,* respectively.

The seemingly trivial chemical difference between a nucleotide and a deoxynucleotide makes an enormous difference in the sensitivity of strings of nucleotides to degrading enzymes, and presumably therefore accounts for the widely divergent purposes for which these molecules are used in nature. Both nucleotides and deoxynucleotides are capable of forming long strings consisting of the chemical union of nucleotides or deoxynucleotides. This process is termed *polymerization* and can result in the formation of deoxynucleotide chains of enormous length. These chains consist of sugar backbones joined by phosphate linkages. From each sugar hangs a specific base. It can be readily appreciated that each chain has an intrinsic asymmetry, having a 5' and a 3' end because the individual nucleotides are joined 5' to 3' through their respective sugars.

The bases themselves are composed of two distinct groups, the pyrimidines (cytosine, thymine, uracil) and the purines (adenine and guanine). Long strings of DNA occur in plants and animals as double helixes—that is, two strands are usually wrapped around each other to form a helix (see figure 2). This double helix structure is maintained by weak (hydrogen) bonds that occur between purines in one chain and pyrimidines in the other. Even more surprising than this helical structure is the fact that not just any purine is found opposite just any pyrimidine in the opposite chain. Rather, adenine is *always* opposite thymine, and guanine is *always* opposite cytosine. The pairing results in part from the basic chemistry of the compounds and also in large measure from the fact that enzymes (pro-

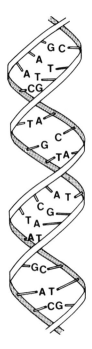

Fig. 2. The double helix of DNA.

tein catalysts) construct DNA on that plan. Thus DNA in higher forms consists of two strands of DNA, each a reflection of the other in the adenine-thymine, guanine-cytosine mirror, with the exception of the fact that the 3' end of one chain finds itself opposite the 5' end of the other. This asymmetry is important, for it is the basis of the fact that the two chains can be distinguished biologically (more on this later).

What about uracil? In fact, this base has little role in DNA but rather is found in RNA (see figure 3). Under certain circumstances, an RNA strand can be produced as a mirror image of a DNA strand, but in this case adenine in the DNA strand is paired with uracil in the RNA strand—thymine not being found in RNA. One could think of uracil as being the RNA equivalent of DNA's thymine. The significance of this substitution of uracil for thymine is not entirely clear.

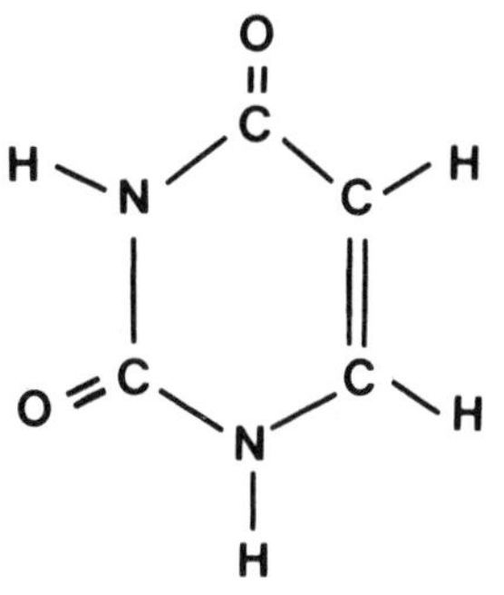

Fig. 3. Molecular structure of uracil.

The story of DNA and RNA as outlined thus far indicates that great complexity can reside in these long nucleic acid molecules—many of which contain millions of base pairs. It is easy to see that a great deal of information could be stored in these molecules—and each string of information seems to be contained in duplicate (because of the double-stranded nature of DNA). But until the rest of the mechanisms of biological control are explored, we are left with simply some interesting chemistry. The key to the process comes with an understanding of protein structure.

Proteins, as noted above, serve as important regulators of cell metabolism. The three classes of proteins most intimately involved in these control processes are enzymes, hormones, and receptors—the latter two of which will be discussed later. Enzymes are catalysts—they facilitate specific chemical reactions. For example, many enzymes together help catalyze the overall reaction $C_6H_{12}O_6 + 6O_2 \rightarrow 6CO_2 + 6H_2O$ + energy—the metabolism of sugar. This reaction would occur to some extent in the absence of enzymes, but at a rate so slow that only negligible quantities of glucose (sugar) would be converted to energy in a reasonable time. The reader can verify this for himself by noting that when he takes the top from his sugar bowl in the morning he is not aware of any spontaneous "burning" of the sugar. Yet in the presence of the enzyme systems of living organisms this energy production is greatly facilitated, and in fact virtually all cells, including those of man, use this reaction, at least in part, to generate sufficient energy to carry out their metabolic functions. How do the enzymes that metabolize glucose—not to mention the thousands of enzymes that do other equally impressive chemical tricks—work? The secret lies in the structure of the proteins.

Proteins, like DNA and RNA, are composed of strings of basic units; in the case of the proteins, these units are amino acids. The proteins found in virtually all living forms are essentially composed of strings made up of the twenty or so different amino acids (see figures 4 and 5). This is not strictly true, because many amino acids are chemically modified once they become part of a larger protein chain. And in fact science every so often discovers a new amino acid in nature. For example, an amino acid in the drug cyclosporin A (a new agent that promises to revolutionize transplant surgery by, in a relatively nontoxic way, suppressing rejection) has recently been found to contain as part of its structure another amino acid that has never before been identified.[6] But for the most part, all proteins consist of twenty basic building blocks. Given the multiplicity of subunits that can be used to make a protein, one can imagine that the electrical and chemical forces in the near vicinity of two different protein chains may be very different. Indeed, different conditions are to be expected alongside different parts of the same protein chain. This can most easily be appreciated when one realizes that, while all free amino acids contain the acid functional group COOH, some contain acids or bases in their side chains. Thus, some remain acids after incorporation into protein chains, while others become neutral or basic, depending on the side group substituents of the individual amino acid. Clearly, it is a matter of some importance to a small molecule whether or not it is located near to a base or to an acid. Additionally, the varying physical chemical forces resulting from the chemistry of the individual amino acids cause many protein chains to interact with themselves, forming complex coiled and wound structures (resulting in so-called tertiary structure). In the

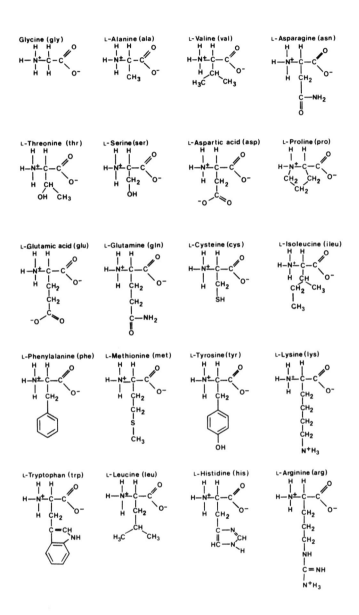

Fig. 4. The twenty amino acids commonly used as building blocks in proteins.

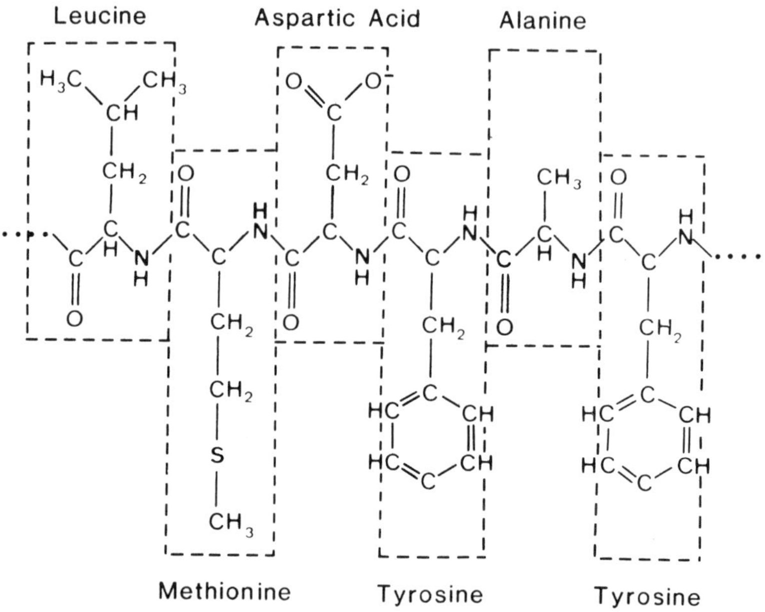

Fig. 5. A representative portion of a typical protein chain. Individual amino acids (shown in boxes) are joined through peptide linkages to form the chain.

case of enzymes, these tertiary structures lead to the generation of pockets composed of amino acids from various parts of the initial chain, into which specific target compounds for the enzyme can fit. In the pocket—or "active site"—the chemical environment facilitates a specific chemical reaction involving the target compound of the enzyme.

For example, if one envisions the combination of glucose with oxygen to yield carbon dioxide, water, and energy as roughly analogous to a marble (glucose) sitting in a large bowl at the top of a hill with a second marble (oxygen) at the bottom of the hill, one can get a fairly good idea of enzyme action. One could argue that the glucose marble, if it were to roll down the hill, would collide with the oxygen marble, shattering both, producing new fragments composed of bits of both former marbles and generating a good deal of

22

energy in the form of motion of the fragments (heat). But the glucose marble is unlikely to roll down the hill because it is trapped in the bowl (that is, the amount of energy required to activate the reaction is large). One could provide heat, thereby causing the glucose marble to move faster and faster in the bowl until it spins out of the container and descends the hill. That this approach will stimulate the reaction is obvious from experience, given the well-known fact that sugar will burn if a match is put to it. But cells don't like extreme heat, so enzymes are employed to provide protected sites in which the energy of activation of one or more reactions is reduced (the bowl is removed) so that specific reactions can proceed at the temperature of living organisms. Additionally, enzymes allow the channeling of the energy produced by reactions. In living cells, for example, the energy derived from glucose metabolism is not lost as heat, but rather is stored for future use as high-energy phosphate (ATP). Enzymes, then, allow reactions permitted by the laws of thermodynamics to occur more efficiently. They serve as catalysts and thus can orchestrate the chemistry of the cell.

It is quite remarkable that the initial sequence of amino acids in an enzyme (its so-called primary sequence or structure) results in folding (tertiary structure) so that a pocket (active site) is formed of just the right configuration to facilitate the joining of specific molecules. This observation becomes even more mind-boggling when one considers that there are thousands of these enzymes in cells performing a tremendous variety of sophisticated chemical tricks. Enzymes exist that excise certain sequences of DNA, uncoil DNA, join nucleic acids and deoxynucleic acids to form RNA and DNA, generate energy (in the form of ATP, the high-energy-containing compound that can donate its energy widely in the cell and so drive the whole mechanism) from the combustion of carbohydrates, fats, and proteins—and do a great deal more. Indeed, it is the exquisite specificity of enzymes that permits the orchestrated activity of the cell to function, for in their absence the chance that the chemical reactions needed by a cell would actually occur based on molecular affinities would be vanishingly small. That they would occur in the proper sequence is even less probable. Enzymes are nothing less than the pro-

23

moters of life, and all these powers derive from the initial amino acid sequence of their primary structure.

The first major advance leading to the present revolution in molecular biology occurred early in this century and derived from the study of human inherited diseases. There exist a large variety of inherited disorders in man, some lethal or extremely debilitating, in which a specific metabolic function cannot be carried out. For example, some people cannot metabolize the amino acid phenylalanine, which they ingest in their diets. This defect results in the elimination in the urine of large quantities of products derived from phenylalanine. This occurs because, in the absence of the normal metabolic pathway for phenylalanine, secondary pathways are flooded with the compound, producing abnormal amounts of some products that accumulate in the blood before being passed through the kidneys and into the urine. Because elevated levels of phenylalanine and its metabolites are toxic to normal brain cells, mental retardation can develop. This disease, phenylketonuria, is therefore characterized by mental retardation as well as by the presence of abnormal levels of phenylalanine metabolites in the urine. Fortunately, medicine has developed techniques for screening the urine of newborns for these products and can prevent the development of mental retardation by excluding phenylalanine from the diet.

Following the intensive study of a variety of such diseases, an important observation was made. Many of these heritable diseases result from the absence throughout the body of a single enzyme (in the case of phenylketonuria, the enzyme is phenylalanine hydroxylase, which metabolizes phenylalanine to tyrosine) and are inherited by simple Mendelian laws of genetics. Thus, if one parent suffers from a heritable disease that is "autosomal dominant," on the average one-half of his children will be affected. If one parent has a disease classified as "autosomal recessive," none of his children will have the disease but all will be carriers. If both parents have a recessively transmitted disease, all of their offspring will have the disease. These heritable diseases follow the rules of genetics laid down by Mendel for his peas. It will be recalled that the Austrian monk Gregor Johann Mendel explained the results of cross-fertilization experiments in peas by assuming that inheritance of any character-istic is transmitted in a lump—later called a gene—which could be-have "dominantly" or "recessively" depending on the specific char-acteristic in question. The scheme works throughout the animal kingdom, including in man. Tay-Sachs disease in man, for example, is recessively inherited. Of course, neither Mendel nor anyone else in his time knew exactly what a gene was, nor was anything known about how these genes worked until the study of inherited metabolic diseases was undertaken. And then the answer began to become clear.

You see, if each of the Mendelianly inherited diseases—diseases presumably caused by an abnormal gene—is indeed associated with a defect in one enzyme, could it not generally be true that one gene controls the production of one enzyme? This concept, which origi-nated with Garrod's work on inherited diseases of metabolism, was explicitly formulated by Beadle and Tatum and further refined by Jacob and Monod. *One gene controls one enzyme.* This idea, al-though not entirely correct, was the midwife to Bioburst.[7]

Of course, one might reasonably ask *how* one gene controls one enzyme. A detailed answer to this apparently simple question lies at the heart of this monograph and will be provided in subsequent chapters. It is sufficient to say that a complete answer cannot be given at present, and in fact expanding the answer to this question underlies most of today's basic biological research. What can be said is that the essence of the process lies in the fact that all the magic worked by enzymes is in the first instance derived from the primary amino acid sequence of a given enzyme. At the same time, DNA is composed of long sequences of nucleotides and, unlike proteins, has the capacity for self-replication. If one gene controls the pro-duction of one enzyme, then it is reasonable to suggest that a given sequence of DNA controls the synthesis of the amino acid sequence of a given enzyme. The link between the sequences of DNA and their related protein sequences is termed the genetic code. Indeed the genetic sequences of DNA determine the sequences of virtually all cellular proteins including enzymes.

If one considers any given DNA sequence and starts with the first nucleotide (thereby determining a reading frame), it turns out that each subsequent group of three nucleotides codes for the presence of one amino acid. Thus, one might expect that the triplet DNA sequence of the gene coding for an enzyme would predict the amino acid structure of the enzyme. And indeed this is partially—but only partially—true.

But one still must ask how the DNA sequences control the production of amino acid sequences of enzymes—and even more poignantly, how the self-replicating DNA sequences came to control the synthesis of the non-self-replicating amino acid sequences of enzymes. The first question can be partially answered. The second will form the basis of subsequent speculation.

First, it must be appreciated that in virtually all living cells essentially the same genetic code is operative. Table 1 demonstrates the essentials of this code. It will be noticed that each DNA triplet (termed a *codon*) codes for a given amino acid, with the exception of a few that are simply punctuation marks used in a later stage of protein synthesis. These "stop" codons tell the protein synthetic mechanism that the end of a message has arrived (more about this later). It should also be noticed that some amino acids are coded for by more than one triplet. That is, the code is degenerate. In no case, however, is there ambiguity in the code. Knowing a triplet sequence, one can predict the amino acid sequence coded for. Knowing an amino acid sequence, however, does not provide exact information about the triplet sequence of DNA because of the redundancy in the code.

What appears to happen when a gene is turned on is that double-stranded DNA in the area of the gene is opened a bit—the strands pulled part—probably by a series of enzymes, so that another enzyme, RNA polymerase, can attach to one of the strands (the "sense strand") and proceed methodically to make a chain of RNA complementary (*i.e.*, obeying the rules of base pairing) to the gene's DNA. The point at which RNA polymerase begins to copy the DNA is termed an *initiation site,* since it is the spot at which transcription (the formation of an RNA chain complementary to a given DNA chain) is initiated. Thus, a gene beginning TAGTCAGTTGGATC

TABLE 1 *The Genetic Code*

First Position* (5′ End)	Second Position				Third Position (3′ End)
	U	C	A	G	
U	Phe	Ser	Tyr	Cys	U
	Phe	Ser	Tyr	Cys	C
	Leu	Ser	Term**	Term	A
	Leu	Ser	Term	Trp	G
C	Leu	Pro	His	Arg	U
	Leu	Pro	His	Arg	C
	Leu	Pro	Gln	Arg	A
	Leu	Pro	Gln	Arg	G
A	Ileu	Thr	Asn	Ser	U
	Ileu	Thr	Asn	Ser	C
	Ileu	Thr	Lys	Arg	A
	Met	Thr	Lys	Arg	G
G	Val	Ala	Asp	Gly	U
	Val	Ala	Asp	Gly	C
	Val	Ala	Glu	Gly	A
	Val	Ala	Glu	Gly	G

*RNA polymerase makes RNA from 5′ to 3′.
**Chain terminating codons.

would be copied by RNA polymerase to form an RNA sequence beginning AUCAGUCAACCUAG—remembering that in RNA uracil (U) takes the place of thymine (T). It should also be noticed that the RNA product is dependent on the correct strand of DNA, the sense strand, being copied by RNA polymerase, for otherwise the wrong RNA would be synthesized. For example, the DNA strand accompanying the gene segment just mentioned—its so-called nonsense strand—consists of the nucleotides ATCAGTCAACCTAG, and the complementary RNA, were this strand to be transcribed (which it normally is not), would be UAGUCAGUUGGAUC, which is quite different from the correct RNA given by the genetic code (that being AUCAGUCAACCUAG). Thus it is terribly important that the sense strand be identified by RNA polymerase and be transcribed. The nonsense strand seems to be important for the correction of errors in the sense strand and for permitting the replication of two strands of DNA during the process of DNA replication. It is the memory strand, if you will. How RNA polymerase is able to identify the sense strand remains under active study.

The RNA formed from the DNA template is in essence a messenger carrying the sequence information contained in the gene's nucleotide sequence to the machinery responsible for protein synthesis. This "messenger RNA" (mRNA) leaves the DNA gene template—which in the cells of higher forms of life, as opposed to bacteria, is enclosed in a membrane-bound sac (the nucleus)—and travels to localized proteinaceous structures in the cell called ribosomes. Here at the ribosomes a new player appears, the one holding the key to the genetic code. This player is a long-tangled bit of RNA called "transfer RNA" (tRNA). In many ways it is similar to enzymes in that it has a complex tertiary structure complete with two active sites. One site consists simply of an exposed triplet codon of RNA; the second is a hospitable environment for a specific amino acid—indeed, for *the* specific amino acid for which the triplet sequence of the first site codes.

The magic of this so-called transfer RNA is that it is an RNA that has the key functional capacity (active site) of an enzyme. This is what makes the whole code work. As the ribosome permits triplet codon after triplet codon to be exposed on the messenger RNA, transfer RNAs, complete with their specific amino acids in their active sites, are constantly floating by. When a transfer RNA with a triplet sequence complementary to the exposed messenger RNA codon approaches, it is held in position by the physical chemical forces resulting from nucleotide-based pairing. As a result of this positioning of the transfer RNA, its specific amino acid is necessarily positioned in the proper location along the growing protein chain and, through a not entirely clear process, is fused with that chain. In the process, the transfer RNA is freed from both the protein chain and the messenger RNA codon. The ribosome then moves over one codon on the messenger RNA and the next amino acid is put in place. When the ribosome reaches a stop codon, the protein chain is released and one new protein molecule has been synthesized.

The process is repeated again and again, depending on the number of messenger RNA molecules that were initially transcribed from DNA, the average lifespan of these messages before enzymes in the cell destroyed them, and a host of other regulatory factors present in the cell. Indeed, what has been outlined here is only the bare bones of the process. Many regulatory and facilitory factors affect the "transcription" of DNA into RNA as well as the "translation" of messenger RNA information into protein, but the essence of the process is as we have outlined it. Information in the DNA of genes is converted into an RNA message (complementary to the DNA of the gene), which then is translated into protein. By such mechanisms are generated all the myriad of primary protein sequences which inexorably, by the laws of chemistry and physics, fold themselves into thousands of life-sustaining enzymes.[8]

Before going into the complexities and ramifications of this scheme, it might be worth attempting to answer the second question we posed—how did DNA nucleotide sequences come to control the production of protein amino acid sequences? No one knows for certain, so any attempt to answer must lie somewhere in an area between hypothesis and fantasy. Still, it is likely that the primacy of nucleic acids over amino acids derives from two facts. First, nucleotides can, by base pairing, form multiple copies of themselves —permitting both self-replication and signaling by messenger RNA production—but amino acids by chemical happenstance have been unable to work this trick efficiently. Second, RNA has the capacity both to read the entire DNA code and to form tertiary structures capable of interacting like enzymes with amino acids. Indeed, recent evidence suggests that some RNA sequences can catalyze simple reactions involving RNA itself. Amino acid chains can recognize segments of DNA and potentially could subserve the function of transfer RNA, but they have never bothered to do this, and probably with good reason. Lacking any readily available means of self-replication, the amino acids would always be dependent on the nucleic acids for procreation. So why should they not use them for the messenger function as well?

With these ideas in mind, we turn now to a hypothesis-fantasy.

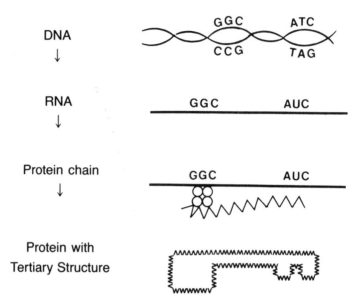

Fig. 6. Transformation of DNA information into protein.

Technical Summary for Chapter II

A.

1. The blueprint for all the tools and machinery used by cells is contained in DNA. The message of this blueprint is written in three-letter words, each made up from the chemical letters A, T, C, G. The DNA message that codes for any given cellular tool (protein) is called a *gene*.

2. DNA can send a chemical mirror image of itself (mRNA) to the workshops (ribosomes) of the cell.

3. At the cell's workshops, the mRNA directs the construction of the new protein—a long chemical made up of different combinations of twenty distinct amino acids. Each of the three-letter words in DNA directs the addition of a specific amino acid to the protein chain.

4. When complete, the amino acids of the protein naturally fold themselves to form useful tools for the cell. Thus, a given sequence of DNA directs the production of a specific tool needed by the cell.

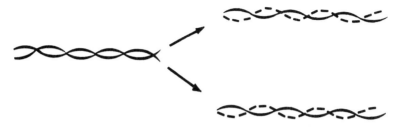

Fig. 7. Replication of DNA strand into two strands.

B.

1. Because DNA in cells always consists of one DNA strand bound to its chemical mirror image (complementary strand), when a cell divides to make two cells, the two strands of DNA separate, each making a chemical mirror image copy of itself, with the result that each of the two newly formed cells has a complete double-stranded blueprint. Thus, the process can be repeated indefinitely without loss of information.

31

CHAPTER III

Aside:
Fantasy or Speculation

In some ancient time in the earth's past, organic molecules formed from the continuous input of solar energy into the gaseous environment of earth began their long march toward life. Simple organic molecules, to be sure, had already formed in the gas clouds between the stars, but it was on earth that more complicated molecules could accumulate and interact in the protected environment of water oceans. It is not inconceivable that chains of ribonucleotides formed with the help of primitive inorganic catalysts in the inlets of the earth's seas, only to be continually destroyed by heat or chemical energy. Nor is it inconceivable that some RNA chains by chance demonstrated enough base pairing between the first and second halves of their length to bend over and bind into tight hairpin shapes. These molecules, we know, are compact and incredibly resistant to heat and ionizing radiation, so they survived while other chains disintegrated. Ever so rarely, one of these hairpins might have found itself in a somewhat more hospitable environment, perhaps in a crevice of a small rock, in which the ionic conditions of the water favored at least temporarily the relaxation of the binding between paired bases; therefore, the molecule's bend opened. Complementary RNA nucleotides could then sufficiently align themselves with some of the bases to allow spontaneous linking to occur, thereby producing a new complementary RNA strand. The temperature in the protected environment of the rock may have increased a bit with the rising of the sun, freeing both strands. These RNA chains might then have drifted out of the protected environment, whereupon conditions favored reestablishment of the hairpin shape. The varying in-flux of energy and changes in the ionic strength of the water that orchestrated this scenario could have been related to daily variations in sunlight, temperature, and tides. It is also possible that some primitive enzymatic activity intrinsic to the RNA chains themselves could have catalyzed not only the formation of new chains but also the exchange of sequences between existing chains. In any event, the primitive viroid replicated, reproducing itself in every other gen-eration, the intervening form being its complementary strand. The base pairing that protected the original viroid also protected its complementary strand so that, with time, the number of viroids increased.[1]

This process occurred many times, producing many different vi-roids that differed in nucleotide sequences and length. For a long while, life consisted entirely of viroids swapping bits and pieces of RNA with their neighbors. An entire society of viroids grew up with genetic inheritance, offspring resembling themselves, and increas-ing population—all the while waiting for the next big step forward.

And that advance came with a bolt. At some point, some of the longer RNA viroids developed enough folding to create active sites for amino acids, and the transfer RNAs (tRNA) were born. With this, some viroids suddenly developed a new mode of action when in a protected environment. They could either reproduce themselves by attracting ribonucleic acids, or they could randomly produce small bits of protein by attracting nearby transfer RNAs with the subsequent spontaneous formation of peptide bonds between the juxtaposed amino acids. For their part, the transfer RNAs had two modes of action. First, in an extremely favorable environment, they could open their tertiary structures and replicate. Second, they found themselves at times involved in the formation of proteins on viroid templates. Of course, not all transfer RNAs used the same RNA–amino acid code that we have today; indeed, a large number of such codes may have existed at the same time. In any case, by this time two distinct species were extant—viroids and transfer RNA.[2]

The first primitive cell was born when a viroid found itself in a protected environment with several transfer RNAs in such circum-stances that, when the transfer RNAs lined up according to the viroid's code, the protein that was subsequently formed was capable of polymerizing into something that closely resembled jelly, trapping within it the viroid and the transfer RNA, and eventually floating out into the world. From this point on, the chemistry of the gel (protein) began to influence the lives of the viroids and the transfer RNAs in dramatic ways. Some of these gels, depending on temperature and chemical conditions, permitted at various times the unbinding of the viroid hairpins of the transfer RNAs and the subsequent replication of these forms. At other times and under other conditions, the gels became permeable to amino acids, leading to the formation of more gel protein. In more advanced forms, still other proteins were produced, some of which proved helpful as enzymes by facilitating either replication or the protein synthetic process. From time to time, the gels tore, sending two daughter elements adrift. Later, these simple cells also made proteins that enzymatically dissolved pieces of the gel in order to hasten the separation of daughter forms. This habit was, of course, of great survival value to these primitive life forms, since those forms that divided quickly had a better chance to survive.

Then came the next step. Enzyme systems developed to make DNA from the original RNA. This was a great advance, for DNA is more stable than RNA. Survival was again enhanced. The design ran: DNA generated RNA, which then interacted with tRNA to make protein. Now the advances came fast. Enzymes produced not only water-protein gels (cytoplasm) but also, at the interface of the gel with the outside world, they aided in the construction of rigid cell walls. Double-stranded DNA came into being, and the process of DNA replication was refined more and more. But the underlying structure of the viroid remained, with long runs of repetitive DNA for tight base pairing as well as palindromic sequences capable of forming hairpins.

At about this time the DNA code we know today was firmly established. Of all the viroids initially formed on the primitive earth, only a few made it to the point of the DNA-containing cell, and therefore only a very few codes persisted. And these differences soon tended to blur by virtue of genetic recombination. The early viroids likely had exchanged genetic information fairly liberally, and this process continued as cells developed. Genetic exchange continued because it allowed for controlled diversity that protected the cell line from quick extinction when some minor environmental change occurred. Recombination provided an advantage primarily to strands with similar genetic codes, for recombinant events involving organisms using widely differing codes would have led to protein synthetic chaos. Thus, the survival advantages associated with genetic mixing and recombination also tended to focus the literature of life into one language.[3]

The next step occurred when certain cells developed the capacity to use the light of the sun to "fix CO_2." That is, they developed enzyme systems to catalyze the reaction $6CO_2 + 6H_2O + \text{sunlight} \rightarrow C_6H_{12}O_6$ (sugar) $+ 6O_2$. The magic compound that produced this effect is chlorophyll, which is neither a protein nor a nucleotide but represents a class of molecules that is of critical importance to life. These we will deem the *facilitators*—strangely shaped molecules that may initially have been the physical chemical facilitators of various chemical reactions in the viroid and early cell. These come in many forms, most of which are still unknown to us but which may persist in one form or another in the nuclei of our cells. One of these facilitators developed into a catalyst for the CO_2 fixation reaction, and the light of the sun was thus trapped to make sugars. Other cells then developed enzymes capable of partially metabolizing these sugars by fermentation, then storing the metabolic energy so derived as the energy-rich compound ATP, which by this time was driving most of the cell's energy-requiring machinery. No longer were cells dependent on an input of heat energy for division as the viroids had been. Now the light of the sun was systematically collected by chloroplasts, sugars were made and converted into ATP, and cell replication became chemically powered. Still other cells developed even more efficient methods of deriving chemical energy from glucose by utilizing more of the formerly facilitating molecules to produce a system of aerobic generation of energy from the products of fermentation. Oxygen began to appear on earth in increasing abundance, in large part as a product of chlorophyll-induced photosynthesis (see equation above). Some cells developed methods of combining this oxygen with the partially metabolized products of fermentation to release even more of the energy stored in sugars and to generate greater quantities of ATP.

Then a strange thing happened. The chlorophyll-containing photosynthesizing bacteria gave up their free-living ways to take up residence in the protected interiors of other cells. So, too, did the aerobic energy producers. Perhaps they were simply surrounded by other cells and proved to be so helpful to those cell interiors that did not enzymatically digest them on the spot that they and their host cells now flourished together. Although only certain plant cells gave a home to chloroplasts, soon all higher life forms contained the energy-producing cells, which in turn became totally dependent on their hosts. These little energy-producing cells could divide in their host cells and live relatively freely, but their genes soon were no longer capable of self-replication. They required materials coded for by the host genes—after all, why duplicate everything? And so these cells, these organisms, became organelles, the mitochondria, and the higher life forms on the planet thereby developed a rich new source of energy—and a dependence on oxygen for respiration.

So a more sophisticated cell was born. Presumably still more symbiosis between living cells led to the development of an organelle containing the DNA genetic information (the nucleus) that found itself in an exterior cell containing the machinery for cellular metabolism (the cytoplasm). This then became the eukaryotic cell, and it went on to evolve into higher plants, animals, and even man.

Thus, when we look at a man, we see first an integrated whole and then a well-functioning machine consisting of harmoniously operating organs, and then we see that each organ is a colony of individual cells and each cell a symbiotic product of simpler cells, and that each organelle consists of facilitators and genetic material reminiscent of that first stable RNA chain that formed in the protective crevice of a rock eons ago. At each level there likely are innumerable traces of the progression from molecule to man. These ancient chemicals and structures may be to biology what 3°K background radiation is to astronomy. Just as the latter provides insight into the "Big Bang" and the early history of the universe, so the former may offer clues about the origins of life on earth. But the biological remnants of early life, unlike background electromagnetic radiation, probably are not vestigial—they may be active in the cells of today and may be extremely potent. Certainly, it will be difficult for us to identify them and decipher their functions in the normal living things of today, but one has only to consider how the lowly *kuru* agent (whatever it actually is) can deprive a man of his mind and life to wonder about the hidden power they may possess.[4]

CHAPTER IV

Dancing Genes

Although a revolution has been developing in biology since the late 1950s, nothing has so brought it to the attention of the public as the recombinant DNA technology of the 1980s. Fortunes have been made and lost by entrepreneurs attempting to capitalize on its promise. In some quarters, suspicion and outrage have developed at the thought of its widespread application. In others, hope has spread with the prospect of new and abundant drugs. And yet the true impact of recombinant DNA technology has escaped the popular imagination. The fact is that its incalculable promise derives not from the specific pharmaceuticals it can produce or the particular diseases it may help cure, but rather from its capacity to decipher the blueprints for cellular control as they exist now and perhaps as they existed in the past. Therein lies the essence of the new biology.

To see this, recall the year 1954 when the Salk polio vaccine was introduced into general use in the United States. It is hard now to envision the impact of this agent on the physical health and mental well-being of the American people. Gone was the fear of the summer cold, of public swimming pools in July, of great epidemics. The vaccine was based on solid medical principles dating back to the variolation technique of 1721. First, polio virus was chemically treated so as to destroy its infectivity. Its simple components were so badly brutalized that the virus was effectively killed and could no longer cause disease. It was then injected into children, who developed an immune response to the altered virus so that the antibodies formed also interacted with, and inactivated, the natural virus. The power to prevent polio was thus added to mankind's medical arsenal.

But what were these mysterious antibodies? Where did they come from, and what was their nature? Antibodies are large protein molecules made by a class of white blood cells called B lymphocytes. They come in five different classes, and although the specific functions of the five differ slightly, the principles of their action are similar. Each antibody molecule consists of four protein chains joined chemically by disulfide linkages. Schematically they may be described as follows: each molecule of antibody is composed of two light chains (L) and two heavy chains (H) linked to form a large molecule (see figure 8). Functionally, however, the antibody molecule can be regarded as consisting of two *Fab* regions and one *Fc* region. Each *Fab* region contains one antigen-combining site. The combining site is in essence a protected domain akin to the active site of an enzyme and consists of parts of both the light chain and the heavy chain. The combining site is specific for the antigen to which the antibody is raised—it will stick to that antigen and only to that (or a very similar) antigen. Since each antibody molecule has two combining sites, it can bind to two antigen molecules. Thus bridging structures consisting of antigen and antibody can form when antigen comes in contact with antibody. These conglomerate structures are easily devoured by white cells, resulting in rapid destruction of antigens. Additionally, the *Fc* fragments of the antibodies trigger other cellular and chemical responses, leading to further antigen destruction along with local inflammation. It is in the *Fc* fragments and their functions that the five classes of antibody differ.[1]

In the case of killed polio virus vaccine, for example, B lymphocytes located in lymph nodes throughout the body encounter the killed virus and respond by transforming themselves into plasma cells each capable of making an antibody specifically directed to one or another viral protein—and making it in quantity. Some of these proteins, or parts thereof, are identical to those in active polio virus. Because once lymphocytes respond to an antigen with an antibody response they are primed to respond more vigorously at any future

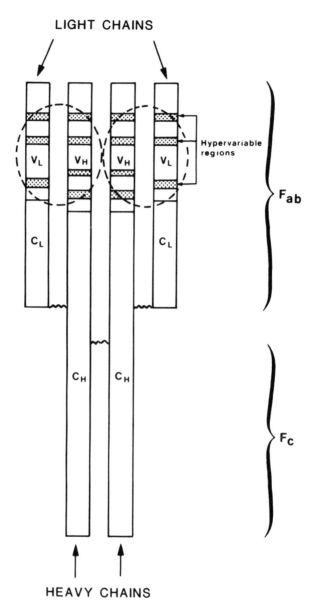

Fig. 8. Schematic diagram representing the structure of a typical antibody. Each molecule contains two antigen binding sites (shown in dashed lines), each of which in turn consists of portions of the heavy and light chains.

time they may again encounter that antigen, these cells obliterate any active polio virus with which they might later come in contact. Immunity results. Thus, although other cells, such as T lymphocytes and macrophages, can facilitate or regulate this process, the production of an antibody-mediated immune response is at heart the work of the B lymphocyte and plasma cell.

The questioning reader will ask, how does the body know how to make an antibody tailor-made to an antigen? Prior to the recent revolution in molecular biology, this was a hotly debated scientific topic.

One theory proposed that the antigen somehow "teaches" the cell to make just the right antibody. This hypothesis recalls the theories of the chevalier de Lamarck, who believed that genetic differences between species were the result of environmental conditions—the giraffe's neck grew long reaching for leaves on high branches. This notion was eventually discredited by the Darwinian concept of natural selection, except in the Soviet Union, where agronomist Trofim D. Lysenko, who based his work on decidedly Lamarckian ideas, controlled the course of genetic research for many years. Lysenko's spectacular failures, the absence of any clearcut example of inheritance based on Lamarckian principles, and most particularly the near-total absence of evidence supporting the theory of antigen-directed antibody synthesis, led to its rejection by the scientific community. For if antigen were to "teach" the cell how to make antibody and if this were not to be accomplished via some Lamarckian genetic mechanism, then antibodies should differ only by virtue of the manner in which they are folded about an antigen template in the cell. And yet antibodies of differing specificities can be shown to differ in their amino acid sequences as well as in their three-dimensional structure. Indeed, if the three-dimensional structure of a specific antibody is destroyed, the antibody can under favorable conditions reform and thereby regain its specificity even in the absence of antigen. Results such as these spelled the doom of the theory of antigen-directed antibody synthesis.[2]

Another theory—indeed, the overwhelming opinion of postwar science—was that antibody production is a process more akin to Darwinian selection than to Lamarckian induction. Because science

41

knew of no established way that an antigen could design an antibody *de novo,* it was assumed that the human animal had in his genes the codes for all the possible antibodies he could make. The one gene –one enzyme (or one protein) law seemed to give the answer. The antigen supposedly triggers those cellular genes coding for the specific antibody that best fits with the antigen.

Objections to this theory soon appeared. First, some scientists doubted that the human genome could contain sufficient information to direct the synthesis of man's entire repertoire of antibodies. (We now know that this is not a valid objection, in large part because the composite nature of the antibody molecule theoretically permits the expansion of the antibody repertoire through the combination of various different heavy and light chains.) There was also the issue of how so many similar antibody genes could have developed and why natural selection permitted them to survive.

In the end a definite answer came from an unexpected quarter. The solution to this problem, as we shall see, was one of the first fruits of Bioburst. But to appreciate this, a bit of background is required.

Research in immunology had by the early 1970s demonstrated that there is a class of white blood cells, the B lymphocytes, that are the progenitors of antibody-secreting plasma cells. These B lymphocytes, as we have noted, are found in great numbers in lymph nodes (the term *lymph gland,* of course, is a misnomer in that the nodes are not major producers of systemic hormones) but also circulate in the blood. It was then discovered that each of these cells contains on its surface copies of the exact antibody that it is programed to make. Should it encounter the antigen against which that antibody is directed—as for example would occur in the case of a developing infection—the binding of antigen to the surface antibody signals the cell to divide and produce still more cells capable of producing the exact same antibody and transforming into plasma cell factories for that antibody. Thus a clone (the progeny of a single cell) is selected so that a large amount of a specific antibody is produced. We can note at once the similarity to the process of natural selection, which is so much in evidence in the grand scheme described by Darwin as well as in more mundane, though terribly significant processes, such as the development of antibiotic resistance in bacteria. Selection of a clone or line with favorable characteristics is one of nature's standard design strategies. And clonal selection explains how a large number of cells making an appropriate antibody can be simultaneously called into play to fight an infection.

But all this information about how large quantities of specific antibodies are produced still begs the question of how the antibodies themselves "acquire" specificity to their assigned antigen. The answer to this question, which had been the bane of immunology for years, arrived with a thunderclap once the techniques of recombinant DNA technology became available. In particular, the use of complementary DNA (cDNA) probes as markers for genes—a process which will be discussed in detail later and which lies at the heart of the new technology—was used to show that, in the fetus, the genes for the various parts of the heavy chains and light chains of antibodies are not located adjacent to one another as the simple schema for transcription that we have discussed would suggest. Rather, they are located at large distances from one another—even on different chromosomes. How can the blueprints for various pieces of linear protein chains such as the heavy chain of antibody exist at such widely separated sites and yet code for one continuous peptide chain?

The answer is, they don't. The young fetus does not make the antibody in question. Rather, with the passage of time, the various gene pieces involved move over long distances to rearrange themselves in the appropriate linear array. *The genes literally jump!* Moreover, multiple genes exist for some antibody regions, and each can jump to combine with any of several junctional and structural genes to form the gene for a unique antibody molecule.[3]

The idea that pieces of human genes routinely jump to new locations flabbergasted the scientific community. (To be sure, Barbara McClintock had shown years before that, in maize, genes could, and did, jump, but somehow this finding was never translated to the human context. But the idea was fundamentally brilliant and eventually brought a Nobel Prize to its originator.)[4] This discovery by Tonegawa, Leder, and their respective colleagues, confirming and expanding the theoretical work on antibody production of Dreyer,

Bennett, and many others, must rank as one of the major medical discoveries of the century. The collection of genes (the genome) of a higher organism had always been assumed to be more or less immutable with any major change associated almost inevitably with dire consequences for survival. Yet it suddenly became clear that whole classes of genes could, as if involved in a giant dance, jump large distances in the normal process of antibody development.

Still more remarkable was the discovery that the jumping genes coding for antibody didn't always jump to the same locations. Even antibody genes formed from the same variable-region and joining-region genes were not the same because gene splicing need not always occur at the same location in all cases. Small differences in the splice locations led to differences in the nucleotide sequences of the newly constructed composite gene. Those differences in turn led to slightly different antibody molecules. As one might guess, many of the splice regions between the jumping genes coded for the hyper-variable regions of the antibody molecule—that is, to the area of the combining sites that determine specificity for antigen binding. Indeed, the binding specificities of antibodies for antigens are in large part determined by the variability of gene jumping, coupled in all likelihood with an increased tendency of genes in the hypervariable region to undergo point mutations in the nucleotide code (with consequent alterations in antibody structure).

To reiterate, it is the composite or modular construction of the antibody gene, the variability in gene splicing at multiple points in the formation of the antibody molecule, and perhaps increased mutagenic instabilities in those splice areas, that account for the vast antibody repertoire of man. Thus, although all cells destined to produce antibody likely derive from fetal cells containing the same DNA composition, they mutate into similar but distinct cell lines during development, each making a slightly different antibody. Each cell makes a single distinctive antibody that is characteristic of it and its progeny. The mechanism of the antibody response becomes clear when one recalls that each B lymphocyte places a copy of the antibody it makes on its cell surface. When the cell is confronted with an antigen to which its cell-surface antibody binds, the combination of the two on the surface triggers cell division so that the total number of B lymphocytes and plasma cells—and therefore the amount of specific antibody made—increases as long as the antigen is present. This is an immune response.

Once the antigen disappears, a residual, relatively large clone of specific B lymphocytes remains, so that with re-exposure to the antigen an even bigger (and in the case of an infectious agent, a protective) response occurs. Thus, an antigen selectively stimulates the production of a clone of cells making specific antibody to it. Because virtually all antigens are chemically complex, a large variety of cell clones are stimulated, each making antibody to one particular characteristic of the antigen and thereby giving rise to the "polyclonal antibody response" that protects us from organisms as diverse as pneumococcus and polio.

It is clear that one of the major questions of twentieth-century medicine—the nature of the immune response—is beginning to yield to the assault of modern biology. The individual genome need not contain at birth the genes for all possible antigens. Rather, individual cells mutate to form a large number of specific antibody-producing clones, the growth of which is then selected for by the presence of antigen. This theory of clonal selection has at least two important ramifications—one theoretical or philosophical, the other eminently practical.

However, before turning to a discussion of these ramifications, it also should be pointed out that the mechanisms of antibody diversity described above may apply to other forms of differentiation in the organism. It may be that jumping genes and selected mutations account for the differing development of cells destined to be brain and those destined to be muscle or skin. In essence, the jumping gene phenomenon and molecular biology may give us the answer not only to the development of antibody diversity but also to differentiation in the organism as a whole. This is an exciting prospect and one that is currently undergoing active investigation, although admittedly little in the way of support for this idea has yet been found. Further, the idea that mutation is a normal facet of development is intriguing for several reasons. First, it implies that, contrary to standard teaching, not all cells in the body contain exactly the same genetic information. Second, an involvement of mutation in development coupled with the known association of mutation and cancer suggests a possible link between development and cancer (more about this later).

Of the two direct ramifications of gene jumping and clonal selection, the more practical application will be taken up first. There exists a spontaneous malignancy in man and animals known as multiple myeloma. This is a cancer, or uncontrolled growth, of antibody-producing cells. These cells grow to fill and replace the normal bone marrow as well as other organs, producing anemia and other dire effects in their victims. Through a complex, poorly understood process, they lessen the production of antibody by normal cells, thereby reducing resistance to infection. Finally, these diffuse cancers produce enormous amounts of antibody—enough in some cases to clog the kidneys and increase the viscosity of the blood, causing as a result sludging of the blood. Without treatment, the prognosis is dismal. Yet this negative prognosis may change dramatically in the next few years, and the myeloma cell itself may prove responsible.

This prospect for the cure of a lethal cancer derives from the fact that the antibody produced in multiple myeloma is homogeneous. The large quantity of protein produced by these tumors is composed of a vast number of identical antibody molecules. In other words, the tumor appears to be clonal. One clone of antibody-producing cells becomes cancerous, producing enormous amounts of a single antibody. This fact was brilliantly exploited by Kohler and Milstein, working in Britain. The contribution of these scientists can be appreciated by considering a simplified version of their ingenious experiments. A mouse was immunized with antigen, and after an appropriate period the mouse's spleen, containing many of its antibody-producing cells, was removed. The spleen was chemically digested to separate it into its individual cells, each of which, of course, went on busily making its own unique antibody against part of the immunizing antigen. The investigators then took advantage of the fact that mouse myeloma cells (obtained from mice suffering from multiple myeloma), like all cancer cells (but apparently no normal human cells), can live indefinitely in tissue culture. At the same time, these cells secrete large amounts of a single antibody to some unknown antigen. Kohler and Milstein, however, reasoned that a form of myeloma cell that secreted little or no intact antibody could presumably still provide a reasonably hospitable environment for antibody production should the appropriate genetic information be introduced into the cell. They wondered if they could introduce the genetic information contained in the antibody-producing cells of their immunized mouse into the hospitable environment of the immortal myeloma cells. In order to do this, they mixed normal immunized cells with myeloma cells and then added a virus known to result, by virtue of its effects on cell membranes, in the fusion of two adjacent cells into one hybrid cell. Thus, normal immunized spleen cells and myeloma cells were fused to produce *hybridomas*—each of which is a very unnatural kind of cell consisting, at least initially, of two of everything. The hybridoma is perhaps best described as the Siamese twins of the cell world.

As one might expect, cell-control mechanisms go a bit haywire when they find another cell living among them, and a variety of chemical reactions take place—including in this instance the persistent activation of antibody synthesis by the genes derived from the normal spleen cells. This persistent turning-on of antibody synthesis produces some remarkable consequences. First, when everything settles down, one is left with a cancerous (read "immortal") myeloma cell producing a single antibody directed at one facet of the antigen with which the normal mouse had been immunized. Since each immunized normal mouse makes many antibodies directed to different aspects of the antigen (a polyclonal response), it is necessary, if one wishes to isolate a single specific antibody, to dilute the newly formed hybridomas sufficiently so that they can be placed one to a well on a tissue culture plate. These cancerous clones soon grow in the wells to the point that they can be injected into the bellies of healthy mice, where they produce enormous quantities of that specific antibody. Scientists need only puncture the abdomen of the recipient mouse with a needle to remove large amounts of antibody. From time to time, cells from the belly of one mouse can be passed to other mice, whereupon these cancerous cells continue their growth in their new hosts; antibody production can in this way be continued indefinitely. Thus, it is possible by these techniques to derive, from the polyclonal response of an animal to an antigen, an inexhaustible supply of many monoclonal antibodies directed to various facets of the antigen.[5]

What are the implications of this phenomenal achievement? One area where it will have great impact is in radioimmunoassay, a widely used technique for measuring the quantities of hormones and drugs that circulate in the blood of man, and one of the cornerstones of modern medicine. Through radioimmunoassay, the diagnosis of pregnancy can now routinely be made as early as two weeks after a missed period by using a radioimmunoassay for the hormone beta-HCG. Acromegaly, hypothyroidism, digitalis toxicity, and a host of other disorders can all be diagnosed using similar procedures. At the heart of the radioimmunoassay method lies the availability of specific antibodies for the hormone or drug to be measured. Traditionally, such antisera are produced by injecting a large number of animals (usually rabbits or goats) with antigen and then testing the animals' blood for the requisite antibody. For an antiserum to be useful in radioimmunoassay, it must bind tightly (that is, with high affinity) and specifically to the hormone or drug to be measured. Many animals, even after months of immunization, can't produce usable antisera. If an animal does make a good antiserum, it is bled repeatedly over its life, but once it dies, the supply of antiserum is at an end. To be sure, the serum of such an animal can be diluted as much as 10,000-fold and still be effective in an assay—such is the power of the antibody response. Nonetheless, the commercial use of antisera can exhaust the available supply in a relatively short time. And there is no guarantee that the next animal immunized will produce quite as good an antiserum. This leads to considerable variability over time in radioimmunoassay results. The inexhaustibility of the hybridoma obviates this problem.

Additionally, in the traditional assay method, an animal-produced polyclonal antibody is used. Some components of this antibody response may bind not only with the hormone of interest but also with antigenic determinants shared by normal body constituents unrelated to the hormone. This leads to the problem of "cross-reactivity." If sufficient quality control is not exercised, cross-reactivity can produce misleading results in a radioimmunoassay. For example, if an antiserum raised to beta-HCG, a product formed by the placenta and therefore diagnostic of pregnancy, should happen to contain one or more antibodies that cross-react with the chemically related hormone LH, which is found in all normal women, the misdiagnosis of pregnancy could easily occur. A great deal of quality-control testing is needed on each batch of antiserum to avoid this kind of error. In the case of monoclonals, the testing need be done only once (or at least relatively infrequently). Also, when a good non-cross-reacting antibody is found, it can be used indefinitely. Thus monoclonal technology promises to revolutionize the multimillion-dollar radioimmunoassay industry.[6]

But that is only a small part of the promise of this technology. For years, scientists have wondered if cancer cells contain specific tumor antigens, or even normal cellular antigens in altered amounts, so that immune sera could be used to destroy them. All attempts in this area met with failure, in that no tumor-specific antisera could be developed. It was unclear, however, if this was because there were no tumor-specific antigens or because the polyclonal antibody response to tumors contained antibodies that cross-reacted with normal tissues, thereby masking any tumor-specific antibodies. Monoclonal antibody technology promises to answer this question, with important therapeutic implications.

For example, consider the case of cancers of antibody-producing cells. We now know, in large part thanks to monoclonal antibody technology, that many cancers of lymphoid cells, such as acute lymphocytic leukemia, non–Hodgkin's lymphoma, and multiple myeloma, are cancers of antibody-producing cells "frozen" cancerously at one or another stage of development. As antibody-producing cells, the cells of many of these tumors contain a specific antibody molecule on their surfaces. It is theoretically possible to produce monoclonal antibodies to the cancer-cell-bound immunoglobulin (*i.e.,* to the antibody on the cell surface). Thus the man-made antibody could selectively bind to the cancer cells and either kill them directly upon binding or be used to carry chemical toxins specifically to the tumor cells. Indeed, one patient with a refractory lymphoma has been placed into long-term remission by the infusion of monoclonal antibodies to specific cell-membrane-bound antigen. In this historic case—a man treated by Dr. Donald Levy's team at Stanford University—the tumor was of B-cell origin and its cells had a specific antibody marker on their surface. But the principle may be applicable to cancers of a different class of lymphocyte, the T cells, as well as to other cancers (initial studies of monoclonals against T cells have not as yet been very successful, perhaps for technical reasons). In any event, the B-cell neoplasms of man that are associated with immunoglobulin surface markers, as many of them are, certainly appear to be prime targets for monoclonal antibody therapy, both because of the terrible toll they produce in terms of human suffering and because of their identifiable surface markers. A world in which acute lymphocytic leukemia is as banal and unremarkable as pernicious anemia or polio is but one of the benefits that could flow from modern molecular biology.

But what of other cancers? Does this technology hold the promise of either new radioimmunoassays for the early detection of cancer or new therapeutic agents for cancer treatment? The answer to this question critically depends on the existence of tumor-specific, or relatively tumor-specific, antigens. Although this point cannot now be determined with certainty, it appears likely from data currently available that antigens with at least relative tumor specificity exist for many kinds of cancer. Already relatively specific monoclonal antibodies have been coupled with radioactive isotopes and used to locate metastatic tumors in patients. Thus, the distant spread of cancer can be determined using these probes. This observation of tumor-related antigen of sufficient specificity to allow the imaging of metastases in turn opens the door for a whole new field of cancer therapy in which cell toxins, such as radioactive materials or non-radioactive toxins like the drug ricin, are coupled to one or more monoclonal antibodies in order to enhance delivery to tumor sites. Ricin is a particularly interesting agent that stops the production of RNA in cells. One molecule appears capable of killing a cell. This material gained considerable notoriety when it was suggested that it had been used by an Eastern-bloc intelligence agency to kill a defector in Britain, the ricin apparently being introduced into the victim's system via the poisoned tip of an umbrella. It would be ironic if this notorious chemical were to help revolutionize cancer therapy.[7]

While the existence of relatively specific tumor antigens may be sufficient to provide improved cancer therapy, the existence of true tumor-specific antigens could theoretically revolutionize cancer therapy by making possible the cure of any tumor carrying such a specific antigen (for cure, it will also be required that all tumor cells display the antigen and that the cancerous cells either not lose the specific antigen with time or succumb to therapy before loss of antigen occurs). Although the jury is still out on the existence of suitable tumor-specific antigens, it is clear that monoclonal antibodies offer the prospect of improved cancer therapy and perhaps of cure. This observation is in part responsible for the tremendous financial interest in firms committed to biotechnology. More important, it is the source of a great deal of hope for present and future cancer sufferers.

The philosophical consequences of jumping genes and clonal selection are harder to define accurately than the tangible practical application to monoclonal antibody technology. Perhaps it is simplest to say that the existence of jumping genes changes the way in which we look at ourselves.

If you ride the bus home from work, you cannot help but notice your neighbors. One may be fat, one may be old, one appears healthy, one frail—each a distinct individual whom you view as a whole, as an entity. A doctor may tell you that the big robust fellow has atherosclerosis of his coronary arteries and isn't nearly as healthy as the frail fellow next to him, but you're probably not comfortable thinking of the two in those terms because you are used to thinking of people as intact entities. The doctor for his part is not uncomfortable with the idea that people are made up of component parts, such as the master gland (pituitary) that controls the peripheral glands and so forth. The human body in this view is a well-oiled machine. No matter that a doctor may prefer to deal with the whole person, he is nonetheless compelled by his training to see a collection of physiologic machines around him on the bus as he travels

51

home from work. The cell biologist when confronted with the same scene sees a collection of cells—colonies, really—all living in harmony to produce a supercolony or "nation" called a *man*. In this view, each of the robust man's cells lives its days in a community, its functioning and well-being dependent upon its individual citizens. There is no master gland but rather a free exchange of hormones and nutrients between sharing partners. The pituitary becomes a colony that exports trophic hormones to the periphery and receives glandular hormones in return. The machine has become a nation-state, the unity of which is assured by the commonality of the genetic material that is shared by all participants and is passed through the germ line (sperm and egg) to explorers who launch forth to found the new nation-states of the next generation. In this view, loss of genetic commonality—loss of "self," or loss of "identity"—usually results in something bad, often cancer, sometimes birth defect.

But the jumping genes show us that not all the cells in the same man contain exactly the same genetic material. In the case of immunoglobulin-producing cells, the genetic compositions of neighboring cells are very similar, but definitely distinct. This observation in turn raises the possibility that genes jump in other tissues besides the lymphoid tissues. Thus our nation-state ceases to be ethnically homogeneous, but must be viewed as a pluralistic multi-ethnic society, the components of which for the most part live in remarkable harmony.

Do not think for a moment that these views of man are simply quaint analogies with no practical importance. The gestalt (whole-entity) view of man has certain definite consequences. The smoker who notices no symptoms or change in his feeling of well-being after years of smoking can be lulled into a feeling that his entire being remains well. He neglects those colonies in his bronchi, or in the lining of his arteries, that have taken a beating with the passage of time. Either the mechanistic well-oiled machine view or the colonial view would help him change his ways. Conversely, the individual who is afraid of any radiation or chemical that can produce, in any circumstances, the least change in DNA can perhaps gain some comfort from the fact that the normal genome is dynamic and fluid, and may not be the rigid and consequently brittle thing he imagined.

The fact is that we will in the future not only have to learn a lot more about jumping genes but also assimilate the consequences of a dynamic genome into our health practices and daily lives. Can genes jump too much, thereby causing disease? Can they jump too little, causing abnormalities in development? Can foodstuffs, vitamins, or drugs affect the jumping of genes in subtle ways that affect health and longevity? As yet we do not know, for until very recently we have been unable to look.

Now we can.

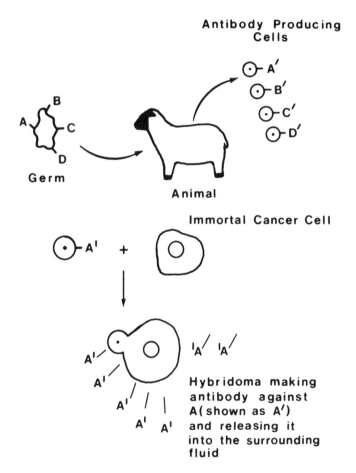

Fig. 9. Development of monoclonal antibodies.

Technical Summary for Chapter IV

A.

1. When a bacterium or virus (germs) or other foreign substance enters an animal, specific cells in the animal react by making chemical bullets (antibodies) against many of the chemicals of which the intruder is composed.

2. If these antibody-producing cells are separated and then fused with an immortal cell line, hybridomas are formed, each of which produces a single (monoclonal) type of antibody.

3. Thus, if a cancer cell should happen to have a chemical X on its surface that is not found on normal cells, monoclonal antibodies can be made against that specific chemical. These antibodies will then not attack normal cells but only the cancer cells. In practice, however, it appears that in most cases only the relative proportion of chemical markers is found to vary between normal and cancer cells, leading to less-than-ideal specificity for monoclonal antibodies. In at least one case, however, absolute specificity seems to have been achieved.

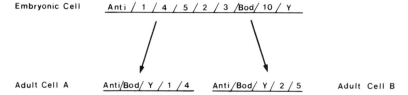

Fig. 10. Role of jumping genes in production of antibody.

B.

1. In the embryo, the genes that code for antibodies are widely separated.

2. In individual antibody-producing cells, these genes jump so as to re-arrange themselves to form the code for a specific antibody. Thus, each antibody-producing cell potentially can produce a different antibody from its neighbor.

CHAPTER V

The Three Keys

Leukemia is perhaps one of man's most dreaded diseases. It also is to chickens, who contend with many of the same ailments as man— including leukemia. In the chicken, however, leukemia is often caused by virus. As early as 1908, it was shown that a cell-free extract derived from leukemic chickens could induce leukemia in healthy chickens following injection. In 1911 Rous reported that the same phenomenon of transmissibility occurred with chicken sarcomas or soft-tissue cancer. In the 1960s scientists discovered that a virus could cause leukemia in cats, and more recently that the same virus can destroy key cells of the immune system in cats, leading to immunodeficiency. Virally transmitted cancers have been confirmed in other animal species as well. These cancer-causing viruses (*oncoviruses*) come in two classes. The first are DNA viruses, and it is not too hard to imagine that injection of foreign DNA into a cell could result in interactions with cellular DNA sufficient to alter cell growth. The second class of oncoviruses, the *retroviruses,* contain RNA, not DNA. It is difficult to imagine how RNA could alter DNA to produce cancer. Yet some RNA viruses clearly do cause cancer.[1]

The full answer to this quandary came with the Nobel Prize-winning work of Temin and Baltimore. The key lay in the existence of an RNA viral enzyme called *reverse transcriptase*. This enzyme is contained in the virus and is injected into the infected cell along with viral RNA. Reverse transcriptase then copies the viral genomic RNA into DNA. It is indeed a "reverse" transcriptase. The DNA copy of viral RNA is usually integrated into the cell's genome where

55

it is slowly transcribed, making viral RNA and proteins. Infected cells often shed virus continually (unlike an ordinary viral infection in which the injected virus runs amuck making new virus, ultimately causing the cell to burst and send forth enormous amounts of virus). In the case of certain viruses, the integrated viral DNA sometimes causes a cancerous transformation in the cell. The mystery of these so-called RNA retroviruses was thus solved.

But that was really only the beginning, because the isolation of reverse transcriptase proved to be one of the three keys which together started the giant engine of recombinant DNA technology. In order to see the significance of reverse transcriptase, we must understand that (a) while every cell contains at most only a few copies of the genes for any given protein (in a genome consisting of thousands of genes), it contains many copies of the messenger RNA for that substance if the cell is indeed synthesizing the protein; (b) although RNA is easily broken apart by common enzymes and so is unstable, DNA is relatively much more stable; and (c) DNA, but not RNA, can replicate coordinately with the genetic material of a cell. Reverse transcriptase then offers the opportunity of using relatively abundant and easily obtained gene transcripts (*i.e.,* messenger RNA) and converting them to a stable, potentially replicating form (DNA). We will return to a further analysis of these properties of reverse transcriptase after briefly discussing the other two keys for the DNA dynamo.

The second key was really provided by clinical medicine and the biological revolution of our fathers' generation—penicillin and the other "wonder drug" antibiotics. These selective bacterial poisons have had enormous impact on the practice of medicine. The tragedy of the otherwise healthy young man dying of pneumococcal pneumonia was obviated by penicillin. Other antibiotics rendered such formerly lethal illnesses as peritonitis and bacterial endocarditis survivable. Tuberculosis no longer required months or years of hospitalization with even then only a prospect of remission offered to the patient. Suddenly, in most cases it could be rapidly cured. It is startling to recall that when penicillin was first developed it was so precious a commodity that the suggestion was made to save the urine of patients treated with it so that the unused penicillin excreted in urine could be purified for reuse. Given the impact of these pharmaceuticals on medicine, it is no wonder that a considerable amount of private and public research was directed toward the design and production of new antibiotics.[2]

In time, however, a problem developed that dimmed the apparently brilliant prospects of antibiotic therapy and raised serious questions for the medical profession. That problem was resistance, which refers to the fact that by random chance one or another subset of microbes in any species of microbes can be expected, through the process of selection, to flourish over time because of a resistance to the effects of a given antibiotic. This resistance can arise, for example, because the resistant organism possesses enzymes capable of degrading the antibiotic or because it has for some reason become unresponsive to the action of the drug. If one were to imagine a billion tubercle bacilli happily chomping away at one microscopic corner of the lung of a person suffering from tuberculosis, one could then consider what might happen if a small fraction of these bugs were resistant to the drug commonly used to treat tuberculosis—isoniazid. Were the patient to be treated with isoniazid, the vast majority of these organisms would be dead in a few days. The remaining resistant organisms, however, might find conditions substantially improved, in part because they no longer had to compete with a billion other organisms for food and space. In time, these resistant organisms would flourish unhampered by competitors eliminated by the antibiotic. If the patient recovers relatively rapidly and the isoniazid is stopped, the resistant organisms could be destroyed by competition with normal bacterial inhabitants of the respiratory tree or by the body's natural defense mechanisms. However, this outcome is not to be counted on. In fact, if a second drug is not routinely administered with isoniazid so as to attack the isoniazid-resistant tubercle bacilli, patients are unlikely to be cured of their disease. Medicine encounters a similar situation in the treatment of gonorrhea. While most gonococci are susceptible to penicillin, the widespread use of this antibiotic has led to the emergence of a penicillin-resistant strain of the organism. In fact, a wave of penicillin-resistant gonorrhea is now sweeping over the world, its likely point of origin Southeast Asia.

57

The art of antibiotic therapy resides in using specific drugs against specific organisms (so that nonpathogenic competitors of pathogenic bacteria are not unnecessarily eliminated) for adequate periods of time to destroy sensitive organisms but not for periods that encourage the emergence of resistant strains. This is a complex issue and one that requires detailed knowledge of the bacterial ecology of the body. In some cases, long-term antibiotics are appropriate; in others, they are not. However, the tendency for antibiotic therapy to select for resistant organisms is one reason why physicians are often reluctant to prescribe antibiotics for colds and other minor illnesses, much to the chagrin of some of their patients. In the case of a cold, for example, the problem is most often caused by a virus. Treatment with antibiotics would do the patient no good but could lead coincidentally to the development of antibiotic-resistant bacteria. This phenomenon of resistance also explains the displeasure of many health officials with countries that permit over-the-counter purchase of antibiotics without prescription. It could be argued that most of these consumer-prescribed antibiotics are misused and have led to the emergence of resistant strains of many organisms. The serious consequences of these developments notwithstanding, it must be noted that the clinical emergence of resistant organisms, while not rare, is not terribly common, and in most cases the development of a new antibiotic has made it possible to eliminate strains resistant to one antibiotic with a second drug. Recognition of this fact provided still greater impetus to pharmaceutical firms to develop new agents, and they responded quite successfully. By the mid 1950s, all appeared under control—in theory.

And then something quite unexpected happened. Some organisms were found that suddenly became simultaneously resistant to multiple antibiotics. That was a startling development from the statistical point of view. It was equally startling from the teleological viewpoint. Why should a bacterium that was treated with penicillin develop resistance not only to penicillin but also to the entirely different antibotic tetracycline, which is chemically quite different from penicillin and kills cells by an entirely distinct mechanism? Did this bacterium, a minute bit of living matter, suddenly deduce from its attack by penicillin that other equally deadly antibiotics were soon to be used against it? Did it suddenly then establish a theory of pharmacology, predict the structure of tetracycline, and take preventive measures? Surely not, for this would be the ultimate Lamarckian nightmare! No teleology fails us completely here, at least at the unicellular level.

The true answer, it turns out, lies in genetics, not teleology. The bacteria that acquire multiple antibiotic resistance carry within their structures small circles of DNA, in addition, of course, to their normal complement of genetic material. These DNA circles, termed *plasmids,* multiply within a bacterium even in the absence of bacterial division. When the bacterium does divide, plasmids usually travel into each daughter cell. (Only certain species of bacteria harbor plasmids; it is for the most part in these species that resistance in general, and multiple resistance in particular, is a major problem.) It is an established, albeit startling fact, that sexual activity exists among certain bacteria. Although these cells for the most part multiply by direct division into two daughter cells, from time to time two bacteria of opposite "sexes" momentarily join their cytoplasms by a long tube of cell sap and exchange genetic material—that is, one injects plasmids into the other. The genes for antibiotic resistance for the most part reside in plasmids. Consequently, if a bacterium sensitive to penicillin and tetracycline sexually conjugates with a bacterium containing a plasmid that by chance carries the resistance genes for penicillin, tetracycline, and ampicillin, the virgin bacterium at once acquires resistance to all three antibiotics. It should be noted that a plasmid containing all three resistance factors will confer on a bacterium a distinct survival advantage in a world in which all three antibiotics are widely (and perhaps indiscriminately) used. In a real sense the widespread use of multiple antibiotics selects for bacteria containing multiple-resistance factors. The teleologist might argue that the plasmid represents the total worldwide experience of all bacteria exposed to all antibiotics. The important point is that for some years now circular bits of DNA, capable of entering and leaving bacteria and capable of multiplying within the bacteria, have been intensively studied. These plasmids are the second key to the engine of recombinant DNA technology.

The third and final key to the recombinant DNA engine is the collection of "housekeeping" enzymes used by bacteria and other organisms to destroy foreign DNA as well as to repair, rearrange, and reproduce their own DNA. There is a large number of such enzymes and probably many more yet to be discovered. For our purposes the *restriction endonucleases* are among the most important. These enzymes cleave strings of DNA whenever specific sequences of nucleotides occur. Generally, these locations are palindromes—that is, the nucleotide sequence of one DNA chain is simply the reverse of that on the other (actually the *same* if read 3' to 5'). For example, the bacterial enzyme Eco RI cleaves DNA whenever the sequence GAATTC occurs. The dashed lines in figure 11 indicate the points at which the enzyme cleaves each of the paired DNA chains. Note that a zigzag cut is made, resulting in a stretch of single-stranded DNA (a "sticky end") at the end of each chain. Note also that the sticky end of one strand of DNA cut by Eco RI will base-pair (hybridize) with the alternate sticky end of *any other* piece of DNA cut by Eco RI, a fact put to good use in genetic engineering (see below, p. 62). Bam HI cleaves at GGATCC and so on. Some restriction endonucleases do not form sticky ends because they cleave symmetrically. For example, Hae III cleaves at GGCC. The importance of these enzymes is that they permit the establishment of landmarks in DNA so that, for example, one can speak of a gene 2 kilobases (2,000 bases) upstream from the Eco RI site in the plasmid pBR322 and expect that the listener knows the location implicitly. Perhaps it is more important that the enzyme opens the DNA at that landmark. Restriction endonucleases are the third key of the engine.[3]

At this point one may ask how these three diverse keys actually start the productive engine functioning. The answer is best found by way of example, because the three keys noted above, used either individually or in pairs, or indeed all together, suggest an enormous number of potentially important and valuable (both intellectually and commercially) applications. A few examples will demonstrate this point.

Assume that one wishes to treat diabetics with human insulin because one feels that it is preferable to the use of beef insulin or because beef insulin is, or will be, in short supply, or because of cost considerations. Can one enlist the help of recombinant DNA tech-

G ⌐AATTC
CTTAA⌐ G

a. Cleavage of Eco RI

G ⌐GATCC
CCTAG⌐ G

b. Cleavage of Bam HI

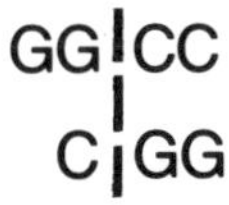

GG⌐CC
C⌐GG

c. Cleavage of Hae III

Fig. 11. Examples of enzyme cleavage.

nology in this effort? The answer is yes—indeed, a human insulin produced by this technology is already on the market. The same technology will soon provide a synthetic growth hormone—a material currently in extremely short supply.[4]

How is this done? Conceptually, one could proceed as follows (see figure 12): the tissue making the hormone of interest or even tumor tissue that is inappropriately (unphysiologically) making the hormone is obtained from a patient during surgery. The tissue is homogenized and its messenger RNA isolated by standard techniques. The RNA is reacted with reverse transcriptase resulting in the production of DNA chains complementary to the messenger RNA. The RNA is then chemically destroyed by treatment with alkali, leaving the DNA. Under these conditions, reverse transcriptase can use the single-stranded DNA it has just made as a template and form a giant loop of DNA, all of which except for the very middle of the loop is double stranded. (Alternatively, DNA polymerase I can be used for making the second strand.) The enzyme Sl-nuclease, which cleaves only single-stranded DNA, is then added and cleaves away the small bit of single-stranded DNA at the very middle of the loop, leaving behind a double-stranded bit of DNA, which codes for the original messenger RNA. Another housekeeping enzyme, *terminal transferase*, is then added, along with a quantity of one or another nucleotide, say dCTP, so that the enzyme adds multiple cytosines to the 3' end of each piece of DNA. In many instances, a prior step of endonuclease digestion leads, depending on the nuclease used, to asymmetric cutting of the DNA strands, resulting in "sticky ends"; in these cases, terminal transferase need not be used since complementary sticky ends would be generated were the plasmid to be used treated with the same endonuclease (see below, p. 66).

This material is then put aside. One next reaches for a standard plasmid, say pBR322, containing one or more known antibiotic-resistant sites, and then reacts it with one or another restriction endonuclease (see figure 13). This opens the plasmid at a specific site. Next, this material is reacted with terminal transferase in the presence of dGTP (that is, a nucleotide that base-pairs with the nucleotide used to treat the hormone gene) so that multiple guanines are added to the 3' end of the opened plasmid DNA. (However, in cases in which the digestion of the hormonal DNA by endonuclease produces sticky ends, terminal transferase treatment of the plasmid need not be used. Rather, the same endonuclease can be employed to produce sticky ends complementary to those on the hormonal gene.)

Finally, the plasmid and the hormone gene are mixed. Because of base pairing, the hormone gene trailing CCC fuses with the open plasmid genes trailing GGG and the plasmid closes with the hormone gene included in it. DNA ligase can then be added to fuse all DNA gaps, leading to the production of a perfectly closed plasmid. This new plasmid is then added to cultures of virgin bacteria (*E. coli* is the bacterium most often used), and in the presence of high extracellular calcium-phosphate concentrations, the plasmid is enveloped and taken up by the *E. coli* (see figure 14). After this *transfection* procedure, the bacteria are routinely cultured, except that an antibiotic is added to the culture medium. The antibiotic chosen is the one whose resistance factor is carried by the plasmid used in the experiment. Thus, only *E. coli* that took up a plasmid grow—and therefore these need not compete with residual virgin bacteria for

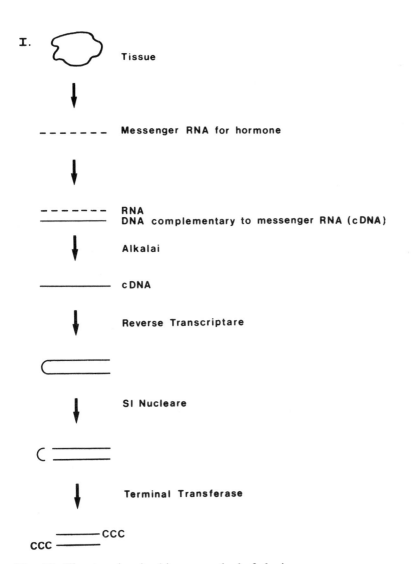

I.

Tissue

Messenger RNA for hormone

RNA
DNA complementary to messenger RNA (cDNA)

Alkalai

cDNA

Reverse Transcriptare

SI Nucleare

Terminal Transferase

CCC

CCC

Fig. 12. The steps involved in one method of cloning a gene:
I. Making cDNA.

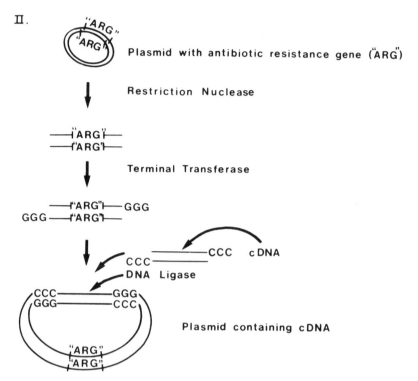

Fig. 13. The steps involved in one method of cloning a gene:
II. Making a cDNA-containing plasmid.

survival. The plasmid-containing bacteria are "selected for" by the antibiotic treatment.

As these plasmid-containing bacteria multiply into huge colonies, many will begin to produce hormone because the plasmid DNA is not only reproduced in the *E. coli* but is also transcribed into RNA during normal cell life—this is to be expected, given the fact that the plasmid-containing bacteria synthesize antibiotic resistance factors. So, too, the cells transformed by a man-made recombinant plasmid will synthesize the hormone coded for by the added cDNA— and they will synthesize it continually and in large amounts. In fact, great vats of cells can be set up to synthesize commercial quantities of protein hormones. Of course, problems remain. The newly synthesized hormone must still be separated from other proteins the bacteria make, and this can be a major undertaking. Nonetheless, purification of recombinant DNA-produced hormone can and has been commercially performed. Insulin, for example, has been commercially synthesized using these techniques.

However, in the case of insulin the process is more complex. In

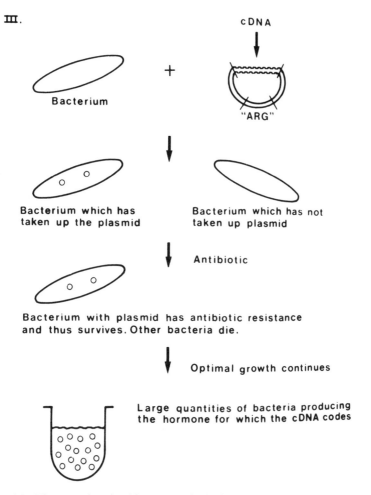

Fig. 14. The steps involved in one method of cloning a gene:
III. Transfecting bacteria.

life, insulin is synthesized as a long precursor that then bends back on itself to form a giant loop. This loop is stabilized by the formation of chemical bonds. Next, the hairpin bend in the loop is excised by enzymes, leaving a 2-chain final product—insulin. Although the entire insulin gene could be cloned, the processing steps needed to generate insulin are too complex to be accomplished easily on a commercial scale. One strategy that has successfully overcome this difficulty involves the separate cloning of gene portions coding for the alpha and beta chains. These protein chains are in turn synthesized separately and then chemically joined to produce insulin. From the insulin example, it can be appreciated that the synthesis of linear hormones is potentially more easily achieved than the recombinant DNA synthesis of multimeric (multiple-chain) proteins, or proteins that are characterized by complex tertiary structures. Nonetheless, with the employment of ingenious strategies a great variety of useful proteins can be synthesized in quantity by these methods.

This description of the cloning of a gene is meant as a conceptual depiction of the process. Many details have been omitted. For example, it is possible to use as a starting material purified hormone message rather than total cellular messenger RNA so as to increase the yield of hormone-producing clones. It is possible to use restriction endonucleases that produce sticky ends, thereby obviating the need for terminal transferase. In the case of small protein hormones (such as the chains of insulin), one need not actually isolate message and use reverse transcriptase to make cDNA. Rather, one can *chemically* synthesize the DNA one nucleotide at a time in the test tube to make the gene of interest—obviously not a process to undertake lightly because of the complex chemistry involved. In the same vein, the use of synthetic DNA linker molecules has greatly simplified the joining of genes to plasmids. As a final example of the complexities of cloning, it can be noted that genes can be cloned in vectors other than plasmids—for example, in viruses. But all these details are relatively minor conceptually. The power of the three keys to ignite the engine of DNA technology is apparent.

Before discussing other applications of this technology, it is important to ponder the implications of the recombinant DNA techniques discussed to this point. On the simplest level, this technology provides an inexhaustible source of scarce protein hormones with obvious medical and financial implications. It should be noted, however, that recently a still newer approach to protein hormone synthesis has been developed that does not involve recombinant DNA techniques, and this process may prove to be the workhorse of the pharmaceutical industry. In this method, hormone-specific messenger RNA is added to ribosomes and appropriate starting materials so that the cell-free synthesis of the hormone can be achieved. Because RNA is unstable, enzymes must be added to produce new copies of the messenger RNA from surviving message and thereby replenish the supply. Although cell-free protein synthesis may be more economical than recombinant techniques because purification of the product is facilitated (after all, one need not separate the product from a myriad of cellular proteins), the scope and applicability of this method are as yet unproven. Recombinant DNA techniques remain the only methods currently used on a commercial scale in the synthesis of peptide hormones.[5]

On a more fundamental plane, recombinant DNA technology is a revolutionary advance. According to classical biology, species are in part defined by their ability to mate sexually and exchange genetic material. The union of cows will result in cows and the union of horses will result in horses. But the union of a cow and a horse can produce nothing. The recombinant DNA technology discussed permits the transfer of genetic material not only across species lines but out of the animal kingdom (man's insulin gene) and into bacteria—a truly remarkable achievement. Although it now appears that certain natural viruses may shuttle bits of genetic material between animals and lower species, recombinant DNA technology for the first time gives man control over this kind of process and presages major possibilities for what has come to be called genetic engineering. The potential impact of this capacity on the health of man and on the social and commercial fabric of his world is hard to overestimate.

A more immediate consequence of the use of the three keys is the development of methods for the sequencing and quantitation of genetic material. This capacity in turn opens the door to a nearly infinite variety of applications in health and commerce. In order to unlock the secrets of the genome and make detailed maps and plans for potential use in the genetic engineering of the future, it is clearly necessary to scout the terrain by determining the exact structure and mechanism of action of genes. Already considerable progress has been made in this area and, in the case of lower forms such as the viruses and plasmids, genomic maps complete with the description of which genes make which proteins are now available. Indeed, progress is occurring so rapidly on all fronts that a basic map (excluding those differences that make for individuality) of the entire human genome could, according to some experts, actually be developed shortly after the year 2000. It is theoretically possible that not long thereafter physicians will be capable of sequencing large stretches of a patient's genome as part of routine medical practice. For example, circulating white blood cells could in theory be taken from infants during well-baby examinations and the genes of these cells analyzed to yield important clues about the risk of the future development of such diseases as diabetes, heart attack, and stroke. In many cases preventive action might be taken to forestall the appearance of these disorders, thereby revolutionizing the practice of pediatrics. But before these potential uses (as well as any potential misuses) of this technology can be fully appreciated, a more detailed understanding of gene-sequencing techniques is required.

Assume for a moment that the gene for a hormone, say insulin, has been successfully cloned. One can then ask what the nucleotide sequence of the gene is. Gene sequence is a terribly useful bit of information, as will become apparent later. In order to answer this and related questions, two methods, that of Maxam and Gilbert, and that of Sanger, have been developed. Although both methods work quite well, we will discuss the Sanger method as the conceptually more straightforward.[6]

The gene is first cloned in a virus so that multiple copies of the gene are available for study. Next, single-stranded bits of the DNA to be sequenced are treated with DNA polymerase in each of four specifically defined chemical mixtures. In the first tube are placed (1) the single-stranded DNA to be sequenced, (2) a small DNA primer to serve as the starting point of a complementary DNA chain,

(3) DNA polymerase, and (4) the four deoxynucleoside triphosphates needed for DNA synthesis—one or another of which is radioactive so as to make radioactive any newly synthesized DNA. Finally, dideoxythymidine triphosphate is added. This nucleotide effectively terminates newly synthesized DNA chains whenever it is incorporated into the DNA because it lacks the hydroxyl group needed for chain elongation. Thus, as the polymerase constructs a new complementary DNA strand, one of two things can happen when thymidine is to be incorporated into the newly synthesized chain. If the polymerase is lucky, a normal thymidine molecule will be inserted and the chain will continue. However, at some point the polymerase will be unlucky and will add dideoxythymidine. The chain will stop. On average, a chain will stop at every point at which a thymidine should have been incorporated. After a period of incubation, the reaction is terminated and the contents of the tube are placed on a gel through which electricity is run so as to separate the molecules in the mixture according to size. This done, a photographic film is positioned over the gel so that the location of the underlying DNA molecules can be detected as a result of their radioactivity interacting with the film. Thus, because all newly synthesized DNA molecules are radioactive and because at least some of the newly synthesized molecules terminate at each location at which thymidine should be found, the photographic film resembles a ladder with a line representing the position of each thymidine. If the same procedure is carried out in additional test tubes in which, respectively, dideoxycytidine triphosphate, dideoxyadenosine triphosphate, and dideoxyguanosine triphosphate are present, the location in the growing chain of these nucleotides can similarly be determined. Because the newly synthesized DNA chains are complementary to the original DNA, this technique achieves the sequencing of DNA.

For example, if the gene contains a sequence such as GCATTAGTA, DNA polymerase will begin a complementary chain CG( ). The third position should be thymidine. If the polymerase is lucky, a normal thymidine will be inserted and the chain will continue CGTAA( ). The sixth position should also be a thymidine. If the polymerase is unlucky, a didt will be inserted and the chain will stop with the following structure: CGTAA(didt), where (didt) represents dideoxythymidine. Since a vast number of complementary chains are produced in the test-tube mixture, on average one or another DNA polymerase can be expected to become unlucky at each thymidine residue in the gene. Thus, the test tube can be expected to contain a mixture of CG(didt), CGTAA(didt), CGTAAT(didt), and so on. If the original DNA treated with DNA polymerase had been derived from cloned DNA by incubation with a restriction endonuclease, then all the DNA treated in each of the tubes begins at a known site, which consequently serves as a reference point. This, in fact, makes sequencing possible. For the purposes of example, let us assume that the gene to be sequenced is, say, four nucleotides (CGCG) away from a known restriction endonuclease cleavage site. Then the fragments generated in our test tube become GCGCCG (didt), GCGCCGTAA(didt), GCGCCGTAATC(didt).

If these bits of DNA are then placed on a gel through which an electric current is passed (gel electrophoresis), they will be separated on the basis of their size. If a bit of photographic film is placed over the gel, several bands will appear on the gel, one band corresponding to each fragment, because the newly synthesized chains are radioactive. This process is called *autoradiography*. Because gel electrophoresis can separate bits of DNA differing by only one nucleotide, the film (autoradiograph) will look as shown in figure 15. In short, the gel reveals the position of every thymidine in relation to the restriction nuclease site. If other copies of endonuclease-generated fragments of the cloned genes are respectively incubated with DNA polymerase in the presence of dideoxyguanosine, dideoxycytidine, and dideoxyadenosine, the locations of these bases can similarly be determined on other gels. The four lanes of autoradiographs when made into a montage would then look as shown in figure 16. Through an analysis of the techniques and vectors used in cloning the gene, the first four nucleotides can be shown to arise from the cloning process itelf, so it can be determined that the cloned gene begins at position 5 with the sequence CGTAATCT.[7]

This technique of gene sequencing is enormously powerful, as the reader might surmise, and one of its immediate but perhaps not so obvious applications is to gene quantitation. In the simplest case, the technique permits one to determine that a gene of interest has

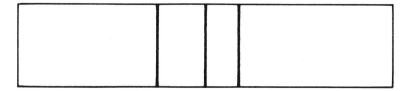

Fig. 15. Autoradiograph of gel electrophoresis of DNA.

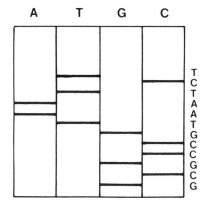

Fig. 16. Montage of autoradiographs of gel electrophoresis of DNA.

indeed been cloned without undertaking the task of determining if the gene's protein product has been made. Second, it provides the structural information necessary to choose restriction endonucleases for the purpose of neatly cleaving the gene out of the plasmid. This is important because cloned genes harvested from plasmids are the major source of cDNA probes used for gene quantitation. Once the technology is available for removing a gene neatly from a plasmid, the gene can then be *nick translated* by specific enzymes—a process that in essence makes it radioactive but otherwise does not affect the gene.

Single-stranded radioactive copies of genes or parts of genes can then be used as probes, thanks to the technique known as "Southern transfer" (named for the developer, not the locale). This procedure takes advantage of the fact that single-stranded DNA will adhere to

71

nitrocellulose paper. Thus, isolated DNA from bacteria, circulating white blood cells of man, or even DNA taken from electrophoresis gels, if it has been separated into two single strands, can be made to adhere to nitrocellulose paper. This done, a probe for one or another gene is placed in a solution over the nitrocellulose paper and, after a period of incubation, the excess probe is washed off the filter. If the DNA sample on the nitrocellulose filter contains the gene to which the probe is complementary, the single-stranded DNA probe will hybridize with the single-stranded DNA, producing a radioactive spot on the paper—which can then be visualized by placing photographic film over the filter paper. The more copies of the gene in the DNA sample, the more probe will hybridize to it and the darker the spot on the film will be. Thus, not only is the presence of the gene confirmed but the amount of the gene is quantitated.

Couple this capacity with the use of restriction endonucleases to clip DNA at known sites and you have the capacity to locate where in the genome a given gene resides. Jumping genes can be located in health and disease. The number of copies of a given gene in a given person can be determined. This in turn has important ramifications. For example, many cancers in man can be shown by these techniques to be characterized by either extra copies or altered locations of a specific class of genes (*oncogenes*). The implications of this finding for cancer prevention, detection, and therapy are, as will be discussed later, potentially enormous. In addition, hybridization techniques have taught us interesting things about how genes work. Resistance to the anticancer drug methotrexate has been shown, at least in part, to be the result of a multiplication of genes coding for the enzyme that methotrexate is designed to inhibit—a previously unsuspected mode of drug resistance. Indeed, the capacity for multiplication in normal cells was a previously unsuspected property of genes, the full significance of which is yet to be determined. Additional insights derived from hybridization study continue to flow in rapidly. The genome of man and other organisms contains stretches of repetitive sequence of DNA—for example, the so-called Alu-sequences—between genes. What function is served by this enormous amount of repetitive DNA can only be hypothesized at the present time. Changes in DNA structure appear to occur with aging and perhaps may lie at the heart of the aging process with senescence resulting perhaps from the gradual clipping out of bits of DNA from the genome, or from inadequate activity of DNA repair enzymes leading to an accumulation of genetic errors. The implications of this kind of observation are very exciting.

By using these techniques in innovative ways, one can in some circumstances diagnose genetic diseases even when the responsible gene is unknown. If by trial and error one finds that (1) a given probe hybridizes to a gene that can be shown using classical genetic family studies to reside near a disease-causing gene, and (2) the abnormality in the disease-causing gene results in the loss of, or the gain of, a restriction endonuclease site, then the length of the fragment cleaved by the restriction endonuclease will differ from normal in the case of the disease-causing gene. The length of the fragments generated in clinical material can be determined by the technique of gel electrophoresis coupled with the use of the available probe to permit visualization of the fragments. This technique, called *restriction fragment-length polymorphism analysis,* has been applied to the prenatal diagnosis of diseases such as Huntington's chorea and cystic fibrosis. It does not provide, in most cases, certainty of diagnosis, in part because it is possible that benign alterations in genes could be present that cause restriction endonuclease sites to change. But in many instances the technique can with high probability point to the presence of an abnormality, and it holds tremendous potential for the detection of human genetic disease.[8]

Much of the major import of this technology will be taken up in later chapters, but there is one other observation that derives from hybridization technology, coupled with electron microscopic visualization of the hybridization of probes to genes, that must be clearly appreciated if genetic mechanisms are to be understood. This is the concept of the *intron.* When mRNA (or cDNA probes made from messenger RNA) were hybridized with total genomic DNA, the probes, of course, hybridized with their parent genes. However, when scientists viewed this process under the electron microscope, instead of finding single-stranded genomic DNA neatly

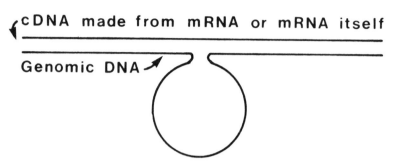

Fig. 17. Intron loop in DNA.

and linearly hybridized (base-paired) with probe, as might be expected, in many cases they noted the strange looplike structures illustrated in figure 17.

These results led to the conclusion that some genes contain within their structure stretches of DNA that are not converted into mature RNA. This was yet another thunderclap in the world of biology because a great deal of previous work in bacteria had suggested that linear contiguous bits of DNA were transcribed into linear strands of RNA, which were then translated into proteins. That this often was not the case in higher organisms was a total surprise. What actually appears to occur is that the extra bits of DNA in the gene loop, termed *introns,* are indeed transcribed into a long precursor molecule (precursor RNA) which is then processed so that the part of the RNA coded for by the intron forms a loop and is removed, never to appear in the final protein product. This is illustrated in figure 18. Base pairing of bits of RNA helped out by enzymes results in the formation of an RNA loop that is then excised. Only non-intron DNA (called therefore *exons*) survive into mature RNA and are subsequently translated into protein.

It is now becoming clear that precursor RNA need not always be processed into the same exons and introns. Depending on what introns are spliced out of RNA, different proteins are made. Different protein hormones, calcitonin, or calcitonin-related protein have been produced from the same precursor RNA as the result of differing intron processing in different body tissues. So, too, the secreted

74

and membrane-bound forms of immunoglobulin M differ in that the membrane-bound form has a long protein tail capable of binding to membrane. Both secreted and membrane-bound forms of this antibody are made from the same gene. RNA processing accounts for the differences in the protein product. There are potentially many other such examples. The undeniable fact is that genetic control occurs not only at the level of the gene but also at the level of the messenger RNA processing. Indeed, mRNA is processed in still other ways. It is often chemically altered at its 5' end, and a tail of adenine residues is often added to the 3' end of messenger RNAs. These modifications presumably affect the efficiency of RNA translation into protein, for it must be recalled that yet another point of genetic control can exist at the ribosome as it synthesizes protein. Anything that alters messenger RNA lifetime or efficiency of translation essentially exerts a genetic influence.[9]

The nature of gene regulation in all its many guises is perhaps the most important question before molecular biology today. Currently we have only intriguing clues to suggest why a gene is turned on and off, either permanently or temporarily, in different body cells, and we have only scratched the surface of determining what other factors can affect transcription and translation. When gene regulation is understood, it will be potentially possible to intervene in such disorders as birth defects, diabetes, cancer, and aging. There is now an

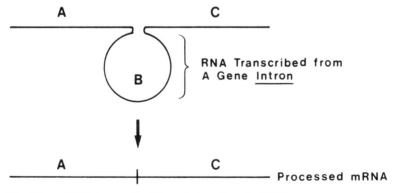

Fig. 18. mRNA transcribed from intron loop in DNA.

enormous potential for medical progress based on the insights already provided by molecular biology. The implications of discoveries about to be made are likely to be still more impressive.

The engine of recombinant DNA technology has started up and is about to shift into second gear.

Technical Summary for Chapter V

A.

1. In cells, DNA consists of two strands, each the chemical mirror image of the other.

Fig. 19. Human DNA growing in bacterium.

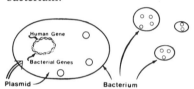

2. DNA pieces of almost any gene (from any living thing) can be grown in large quantities in bacteria. Single-stranded DNA from these bacteria can then be made radioactive, creating a *probe*.

CGTAAG   Radioactive   Probe
————*

+

Fig. 20. Radioactive probe.

3. Probes can be used to detect the presence of their chemical mirror images in any cell or tissue because they will bind to them (*hybridize*) and therefore concentrate radioactivity at the location of the mirror-image DNA.

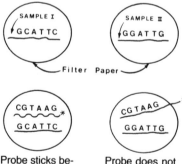

Probe sticks because sequences are complementary

Probe does not stick because sequence is not perfectly complementary. Probe is easily washed off, leaving no radioactive spot.

4. Thus, if the probe is directed against a disease-causing gene (for example, a sickle-cell gene), the probe can be used as a diagnostic test for genetic disease in adults, children, or embryos.

Filter paper therefore becomes radioactive, and this can be detected by the darkening of overlying film.

Therefore gene of interest is shown to be present in sample.

Gene of interest not present *in* sample.

B.

1. When a foreign gene is introduced into bacteria ("cloned in bacteria"), the bacteria make the protein for which the foreign gene codes. This provides, for example, a commercially attractive means of producing for medical use protein hormones that are in short supply.

C.

1. The sequence (*i.e.*, the series of three-letter "words" that comprise the gene) of almost any gene can be determined. Thus, we can now "read" DNA like a book.

CHAPTER VI

Aside:
Fantasy-Prediction

On or about September 2, 1994, a young boy is brought by his distraught parents to a clinic in Houston, Texas. The child represents a tragedy born of poverty and ignorance, not because of the severity of his disease but because of its needlessness. At age twelve, he has been noted in his native Nicaragua to have lost weight, urinated excessively, and developed severe thirst. Without doubt, he suffers from diabetes mellitus. This represents a real tragedy. In 1994, the disease has largely been prevented by the vaccination of susceptible children against the viruses that trigger the body's unfortunate allergy-like attack against the insulin-making cells of the pancreas and lead to diabetes. For those children who either fail to be properly vaccinated or who in spite of vaccination contract one or another of these viruses, immunotherapy can abort the disease in most cases. This treatment, developed in Boston and Bethesda, employs monoclonal antibody infusions to eliminate those clones of immune cells that are triggered by virus, thereby preventing the autoimmune attack from continuing and thus saving residual pancreatic beta cells from extinction.[1]

But the boy from Central America is not a candidate for either therapy. For lack of medical attention, he was never vaccinated and was never brought to a physician for immunotherapy in the early stages of his disease. His pancreas has been ravaged beyond salvage by his own misguided immune system, and all that stands between him and a life of insulin injections—and the likely vascular complications of his disease—is the experimental program underway in Houston.

His first day in the clinic is pleasant enough, consisting of physical examination, history taking, and a discussion of the risks and benefits of the experimental procedures. Despite some last-minute reconsideration and more than a little soul searching, the boy's parents hold fast to their original decision and sign the consent documents. At the end of the day, a dab of antiseptic solution is applied to a spot on the boy's back and a short biopsy needle is inserted through the skin to remove a small plug of skin tissue. One stitch closes the wound, and the boy is told to return to the clinic in five days.

That evening, technicians carefully tease the tissue sample apart in the presence of enzymes that help dissolve the fibrous strands holding cells together. Samples of cells are then placed in multiple tissue-culture dishes with nutrient medium and incubated overnight.

The next morning, a technician removes from a freezer and slowly thaws a small vial of retrovirus X-1. This reagent represents the product of two years' work and consists of a replicative-defective retrovirus containing the human insulin gene and its control region. Based on experiments begun in 1980, retrovirus technology has advanced tremendously. In 1983 human hormonal genes were inserted into retrovirus—mimicking thereby the natural tendency of these viruses to pick up information from one genome and then integrate themselves and their newly acquired genes into a new host. Engineered retroviruses containing human genes were first used to infect rodent cells, with the result that both viral and human hormonal DNA was inserted into the genomic material of the recipient cells. In these experiments, the hormonal genes functioned normally, thereby opening the door for gene therapy at the human level. In essence, the defective retrovirus serves in the human the role that the plasmid serves in bacteria—that of the vehicle for the transfer of genetic material. Over the ensuing years, tremendous strides have been made in designing retroviruses that insert themselves and their newly acquired human genes into safe areas of the genome—areas in which no important cellular function is disturbed by their arrival. Because these viruses are reproductive-defective, they cannot infect other cells and spread their genetic cargo. Rather, only those cells deliberately infected by the scientist with the virus can acquire the new gene. Thus, retrovirus X-l has become recognized as the most specific retrovirus ever produced. It can efficiently and specifically insert its genetic cargo, the insulin gene and its control region, into a safe region in the human genome.[2]

After the virus X-1 thaws, the technician adds fixed aliquots of this material to each dish of the boy's cultured cells. The dishes are then returned to the incubator.

The next day, samples of cells from each tissue well undergo automated analysis. DNA is extracted from each, automatically subjected to multiple endonuclease digestion, separated by gel electrophoresis, and then hybridized against a battery of probes for virtually the entire human genome. The site of insertion of the retrovirus–insulin gene combination is determined in each cell sample. Those cell cultures in which the insertion is found to reside at a dangerous locus are discarded.

On the fifth day the boy returns to have his stitch removed. He is told that after forty-eight hours of continuous automatic screening one clone of his cells has been found in which insertion has occurred at precisely the right spot for proper, safe gene function. He is told to return in three weeks.

Technicians then, with constant care and attention, culture the magic clone. One dish becomes two, two become four, until after three weeks an adequate cell mass has been achieved. Two days are spent testing the product to demonstrate that the genetically engineered cells secrete appropriate quantities of insulin when challenged with glucose. A final mapping of the cells' genome takes place to ascertain that three weeks of tissue culture have not dangerously damaged the cells' normal genetic makeup.

When the boy returns for the second time, he is lightly anesthetized. Under fluoroscopic control a long microthin needle is passed through his abdomen and into his pancreas, the location of insulin-producing cells in normal man. A quantity of engineered cells are then injected. This procedure is repeated a number of times. Afterward, the boy is observed for twenty-four hours to watch for the possible development of pancreatitis as a result of this procedure. When no such sign is found, he is discharged—cured.

CHAPTER VII

Bioburst Versus Cancer, Heart Attack, and Stroke

The two leading killers of Western man are atherosclerotic vascular disease (heart attack, stroke) and cancer. What is surprising to many people is that both of these disorders arise from abnormalities of cell growth. The primacy of a cell-growth disorder in cancer is perhaps obvious. It isn't so clear in the case of heart attack, but in fact both stroke and heart attack are caused by diseases in the arteries to the brain or to the heart, respectively.

The walls of these arteries in some people become scarred and thickened, in large part because of the excessive growth of smooth muscle cells that make up the vascular wall. According to current theory, damage to the thin endothelial cell layer that lines the interior surfaces of all vessels sometimes exposes the underlying arterial smooth muscle cell to blood elements, and then a complex biology develops. Those uninjured parts of the vessels covered with endothelial cells remain free of adherent blood elements largely because these cells synthesize a modified fat called *prostacyclin*—one of the family of compounds termed *prostaglandins*. This prostacyclin in turn prevents *platelets*—small bits of cytoplasm thrown off by giant cells in the blood called *megakaryocytes*—from adhering to the vascular wall. This is important because when platelets adhere to a surface such as the damaged wall of an artery, they coalesce and release a variety of factors, most of which serve to enhance the accumulation of more platelets, thus forming a small plug. This is one of the mechanisms by which bleeding is stemmed following the laceration of a vessel. It can be appreciated, therefore, that if endothelium is removed, for example by the trauma of a severely elevated blood pressure acting on an already damaged artery, platelets will quickly adhere to the underlying tissue. Among the factors released by platelets as they fuse is a peptide hormone called *platelet-derived growth factor* (PDGF). This small protein stimulates the multiplication of underlying smooth muscle cells in the arterial wall and produces a small knot. If the trauma to the endothelial cell layer is transient, endothelial cells regrow over the damaged area and arterial smooth muscle cell proliferation ceases with only a minor bump remaining in the arterial wall to show for the episode. Platelets also carry growth factors other than PDGF, but it appears that PDGF is the most important of these for the development of atherosclerosis. Parenthetically, it can be noted that PDGF is also released when one suffers any minor injury, such as a cut finger. In this case, the growth factor stimulates the multiplication of skin fibroblasts resulting in the formation of a scar. A parallel mechanism operates in the vascular wall, and in a very real sense the fundamental abnormality in the genesis of atherosclerotic vascular disease is the inappropriate formation of a vascular scar.

Although a transient insult to the vascular layer is tolerated relatively well, repeated insults lead to the formation of bigger and bigger knots, which then accumulate fats and other circulating blood cells, leading to the formation of an arterial *plaque* (see figure 21). These plaques are time bombs. If bleeding should occur into one of these, the involved plaque can swell enormously, blocking the lumen of the artery and resulting in the death of all tissue (read "heart muscle" or "brain cells") downstream from that point. Alternatively, if the plaque becomes large enough, the lumen of the vessel may be so narrowed that the natural stickiness of platelets allows them to form a plug (*thrombus*) in the small remaining opening. This again leads to cell death downstream. Attempts have been made to reduce thrombus formation in diseased vessels with drugs such as aspirin, and in select clinical circumstances these have proven to be beneficial. Nonetheless, these drugs do not appear to stop the underlying pathogenetic process of plaque formation.

The scheme just outlined is likely not complete as described. There may be additional growth factors involved, and cell types

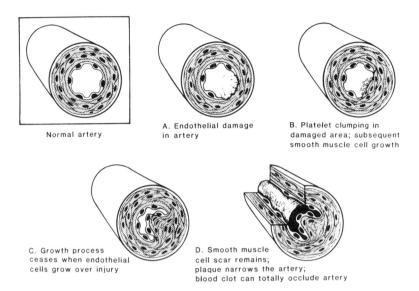

A. Endothelial damage in artery

B. Platelet clumping in damaged area; subsequent smooth muscle cell growth

Normal artery

C. Growth process ceases when endothelial cells grow over injury

D. Smooth muscle cell scar remains; plaque narrows the artery; blood clot can totally occlude artery

Fig. 21. A series of cross-sectional schematic drawings of an artery demonstrating the production of an atherosclerotic plaque with subsequent total occlusion of the vessel.

such as macrophages and endothelial cells almost certainly play important roles in atherogenesis as well. For example, endothelial "injury" need not require the physical removal of endothelial cells from the arterial wall. Rather, endothelial cells themselves may be triggered to release their own growth factor(s), including PDGF, when stressed. Moreover, once in a plaque, arterial smooth muscle cells may produce their own PDGF resulting in some degree of self-stimulation. In some circumstances, such as the presence of high blood-cholesterol levels, macrophages seem to crawl through gaps that develop between endothelial cells to reside among the arterial smooth muscle cells. Here the macrophages accumulate fats and may release PDGF of their own, thereby causing smooth muscle proliferation. This may represent the earliest event in atherogenesis. Actual denudation of endothelial cells and platelet clumping may play a more important role in the progression of the atherosclerotic lesion than in its initiation. Other mechanisms likely are operative as well. But our scheme does represent a theoretical framework for discussion, and it is substantially correct in its own right. This latter point can be appreciated by realizing that atherosclerosis can be dramatically reduced in certain animal models of the disease by drastically lowering platelet levels. The fewer platelets, the less PDGF (and other platelet-carried growth factors). The less PDGF, the less atherogenesis. Although this observation does not in itself have therapeutic significance because the reduction of platelet numbers required to achieve a beneficial effect on atherogenesis is in itself life-threatening due to the tendency for bleeding that it produces and because animal models of atherosclerosis need not fully mirror all facets of the human disease, the experiment does confirm the power of the basic theoretical construct.[1]

If this approach is valid, there are two broad avenues open to science in its attempt to prevent the pathogenesis of atherosclerosis. The first of these involves the elimination of trauma to the vasculature and therefore of the platelet clumping and growth factor release attendant upon that injury. This nostrum translates into lowering blood pressure, reducing elevated cholesterol and glucose levels, attempting to prevent death of arterial smooth muscle cells in plaque (since cell death seems to produce still further injury and reactive cell multiplication), eliminating smoking, and many other beneficial interventions. It can be seen that these therapies in turn fall into two classes: life-style modification (diet, cessation of smoking) and chemical therapy (drugs to lower blood pressure, cholesterol, and glucose). Traditional medicine is directing much effort to this approach and with apparently considerable success. The death rate from cardiovascular disease in the United States has fallen a remarkable 25 percent in the last twenty years or so, most likely as the result of efforts to reduce vascular trauma. And not all potentially useful interventions have as yet even been explored. Recent evidence indicates that drugs known as calcium channel blockers lessen cell death in atherosclerotic plaques and reduce atherogenesis in experimental animal models of the disease. This observation has not yet been extended to man but offers great promise. Prostaglandin analogs may similarly be useful. There is considerable potential remaining in the attack on vascular trauma and platelet aggregation.[2]

However, the major new approach to atherogenesis prevention that is suggested by modern theory is the prevention of arterial smooth muscle cell hyperplasia either by reducing the impact of PDGF and other growth factors on these cells or by decreasing the release of the growth factors themselves following injury. This is a more fundamental approach to the disease and is clearly needed because many individuals are unable to eliminate endothelial trauma even with life-style modifications and currently available drugs—or even with the drugs that will become available in the future. Included in this class of patients may be, for example, people with high blood pressure that is refractory to therapy and people with marked elevations of cholesterol in blood. A direct attack on the atherosclerotic process is required to place these patients beyond risk. It will also be required if medicine is to prevent the still-significant rate of vascular disease in people with no known risk factors for hardening of the arteries. Yet for all its appeal, a fundamental attack on the vascular cell/growth factor interaction is potentially fraught with risk, for unless the cellular biology of the arterial smooth muscle cell is fully understood, the drugs used to influence cell growth might inadvertently exert severely detrimental actions on these cells as well.

Fortunately, science is daily learning more about the cellular and intracellular biology and physiology of the vascular wall. We know that these cells make enzymes and hormones previously thought to be made only in the kidney, and we believe they respond directly to drugs in ways that were previously unanticipated. At least one drug, trepodil, is undergoing extensive study abroad because it seems capable of blocking the effect of PDGF on the arterial smooth muscle cell. It has already been shown to reduce atherogenesis in experimental animal models, and within ten years it, or a similar drug, could be a mainstay of medical therapy. The point seems proven that the cellular biological approach, which is perhaps the single most characteristic feature of Bioburst, appears to be making headway in medical therapeutics.[3]

This last point deserves some reiteration and expansion, because in many ways it captures the essence of Bioburst. In the example of atherogenesis, we see that cell biology offers a whole new approach to this disease, even without using the newer, more powerful tools of recombinant DNA technology and the like. The study of growth factors and other less-heralded aspects of cell biology portends advances perhaps just as powerful as those of the more glamorous areas of the biological revolution—and not just in the arterial wall. Studies of the cell biology of bone, brain, and other organs are making comparable strides, and although they are distinct from recombinant DNA technology, they are all part of the same intellectual movement, the various parts of which continually complement and enhance one another.

It is customary in our day to think of medical advances in terms of technology—better pacemakers, artificial hearts, artificial pancreases, and the like—and although these are part of the cutting edge of modern medicine, they are conceptually distinct from Bioburst. They are, with one exception—that being the union of digital imaging and nuclear magnetic resonance—improvements in technologies derived from the ideas of the past rather than newly emerging, indeed exploding, concepts. This in no way denigrates the importance of these modern technological advances. They are important, and they do save lives. However, although we speak of insulin pump technology and recombinant DNA technology, the use of the word *technology* in the latter case is misleading in that it masks the tremendous intellectual leaps that accompany the use of recombinant DNA and other cell biology techniques. Additionally, the power of Bioburst "technology" to generate new therapies is dramatically greater and wider ranging than many of the conceptually narrow technologies with which we are familiar. Parenthetically, this distinction is also of some importance when one attempts to project commercial trends for the future. For example, success of the Bioburst-derived attack on growth factors coupled with new techniques for cell transplantation may eliminate (or at least substantially reduce) the need for blood pressure medication, and insulin pump technology as well. Other technologies will similarly be reduced in importance. The pharmaceutical industry must, of course, be cognizant of these issues, but so must many other elements of society, including physicians, investors, politicians, educators, sociologists, and theologians.

An additional important point can be made related to current theories of atherogenesis. We have been discussing a collection of diseases that heretofore have been the province of the internist, the cardiologist, the neurologist, or the vascular surgeon. Bioburst in effect is removing heart attack from the exclusive province of the cardiologist and stroke from the province of the neurologist. Either these specialists will have to become adept at thinking in the Bioburst terms of cellular biology or a new breed of medical consultant will necessarily emerge. Some medical disciplines and specialists may become obsolete and either disappear entirely or be forced to refocus their efforts. We see a small foreshadowing of this phenomenon in the case of dentistry, where the potential use of antibiotics and specific vaccines already threatens greatly to reduce the need for dental fillings and periodontal surgery—and consequently to reduce the need for dentists. Medical radiation therapy may be the first of the medical disciplines to come under similar pressure if the Bioburst attack on cancer lives up to its promise. Automated techniques of gene hybridization may replace the anatomic pathologist and his microscope in the diagnosis of many illnesses. The implications of Bioburst on traditional medical practice will be enormous.

Returning to the example of atherosclerosis, it should also be noted that cardiovascular disease is the major killer of Western man and often robs society of the efforts of many of its most productive members at the height of their productivity. The economic consequences attendant upon the maintenance of that productivity into old age are yet another major implication of this technology.

This broad view of the impact of Bioburst on vascular disease makes it clear that the promise of this approach derives from its cellular biological orientation. Hardening of the arteries has been more or less translated into a problem of cell growth and development.

Cancer is also a problem resulting from disordered cell growth and development. Although growth factors may play a role in initiating or maintaining some cancers, they were never thought to play as central a role as, say, PDGF does in the development of vascular disease. On the other hand, there are now hints that these factors may play a much more significant, and possibly critical, role than many suspect. There does not *appear* to be a specific growth factor abnormality causing bowel cancer or leukemia, but then again, there may be. However, the development of cancer appears to derive, at least in part, from an intrinsic abnormality within the affected cell. Thus, the approach to cancer therapy is conceptually more complex than the approach to atherogenesis. (This is a conceptual point only; atherogenesis is an extremely intricate problem when one considers the multiplicity of factors that can initiate or support the process and the cell dynamics that occur in the vascular wall.)

It has often been said that cancer is not a single disease but rather a hundred or more diseases characterized by abnormal cell growth and lumped under the same heading—cancer. There was, and still is, considerable support for this view of cancer heterogeneity. Different cancers have very different characteristics—some are aggressive, lethal, difficult to cure, while others are indolent, responsive to therapy, and associated with near-normal survival. Also, some cancers are associated with one or another age group—Wilm's tumor with children, and rectal cancer with adults. Some are hormonally responsive, others are not. Some spread predominantly by blood and others principally by lymphatic channels. Some occur in smokers while others are more common in certain groups of industrial workers. The heterogeneity of cancer is mind-boggling. And yet at the cell level there are many characteristics common to these disorders. In fact, recent information derived from recombinant DNA–based experiments suggests an amazing genetic similarity between many forms of cancer, raising the possibility that the heterogeneity of cancer may be more apparent than real.

This hope for a unified view of cancer stems not only from the common characteristics that molecular biology reveals in all replicating tissue, but more incredibly from specific observations derived from the study of animal tumor viruses. As has been discussed, tumor viruses such as the Rous sarcoma virus insert into the genome of their target cells a specific gene that is responsible for the transformation to malignancy. In recent years, these so-called viral oncogenes have been cloned, and radioactive probes for the genes have been produced by nick translation. In parallel with studies of tumor viruses, studies of DNA derived from animal and human cancer were undertaken. In these experiments, DNA extracted from human tumors was added to cultures of standard mouse fibroblasts. Some of the added DNA was actually taken up by the normal cells, and some of these apparently normal cells were then transformed into cells that by all criteria were neoplastic or cancerous. They grew indefinitely and lost normal growth-control mechanisms. DNA from human tumors was therefore shown to produce cancer in apparently "normal" cells. In a real sense, the DNA could be termed "infectious." Further study of these transformed cells revealed that only specific genes in the cancer DNA caused the transforming activity. These genes are yet another form of oncogenes, apparently human as opposed to viral. Detailed follow-up studies revealed that the "normal" cells used in these oncogene experiments actually had to be prepared for cancer transformation by yet another oncogene or by a chemical carcinogen and thus were not in a strict sense normal cells. Cancer induction appears therefore to be a process involving at least two steps, with the oncogene-produced steps possibly separated by many years. In theory, therefore, bioscience may soon be able, by the techniques of gene analysis, to determine how far along the road to developing cancer a given individual may be—a prospect of significance to workers in chemical manufacture, nuclear power generation, and other occupations popularly considered at risk for cancer.[4]

Far more amazing than the discovery of oncogenes was the subsequent finding that the transforming genes from human tumors were virtually identical, as determined by hybridization study, with the viral oncogenes previously discovered. Although minor differences existed between the human and animal oncogenes, there was no doubt that close human analogs of the animal-virus oncogenes existed and that these could produce neoplastic transformations in cultured cells. In fact, oncogenes were found in human leukemias as well as in solid tumors such as cancers of the colon, lung, breast, and bladder. In their target cells many of these oncogenes make proteins that have been termed *oncoproteins*. Exactly how these oncogenes and oncoproteins produce cancer is not clearly understood at the moment, but cell biologists are zeroing in on this question. Indeed, not all oncoproteins seem to produce cancer in the same way. To date, approximately twenty oncogenes have been identified, leading to the idea that perhaps only a small, finite number of oncogenes produce most of human cancer—thereby eliminating much of the apparent heterogeneity long thought to be the hallmark of cancer. And with this codification of multiple apparent diseases into neat groups comes the prospect of developing a single intervention that is effective against large numbers of different cancers. One human experimentation committee at a large medical center recently noted that it had on file experimental protocols involving over one hundred sixty drug combinations for the treatment of various cancers. Bioburst-derived insights offer several prospects for the development of a limited number of drugs against a large number of cancers. This is no mean accomplishment in the war against malignancy.

But there is also an ominous side to the cancer oncogene story. If human tumor oncogenes are so similar to virally spread animal oncogenes, is human cancer disease spread by viruses? Infectious cancer is perhaps one of the greatest horrors the mind can imagine. Could an outbreak of infectious cancer spread over the world, killing millions and leading to the quarantine of hundreds of thousands? The world has not had to contemplate this kind of epidemic since the major infectious diseases like smallpox and yellow fever were controlled by public health measures. Infectious cancer could bring back the ethos of the plague years and the Black Death.

On this issue, the news is both good and bad. On the bad side is the mounting, essentially overwhelming evidence that at least one form of human cancer is infectious, and several others probably are. On the good side is the evidence indicating that most human cancers and most human oncogenes are not likely to be virally transmitted. But if oncogenes are not carried by virus, how do they get into the human genome to cause cancer? The answer is perhaps as startling as anything learned to date about cancer. The situation is roughly akin to a man awakening from sleep to find his possessions vandalized—and then discovering irrefutable evidence that in his sleep he himself had done the damage. Hybridization studies have revealed the presence of oncogenes in *normal* human cells. This discovery at first glance appears contradictory: oncogenes cause cancer, and yet oncogenes are present in normal cells. How can this be?

The apparent contradiction disappears when one asks "how many oncogenes are present in normal and in malignant cells?" or "where are the oncogenes located in the genomes of normal and malignant cells?" or "are the oncogenes subtly different from their normal counterparts?" Cancer cells may contain an increased number of oncogenes, or perhaps oncogenes in abnormal locations, or oncogenes that are active when they normally should be turned off, or oncogenes that are subtly altered. It appears that oncogenes are normal cell genes that play important roles in cell growth and differentiation but then are strictly regulated as the cells reach maturity. Relocation of these genes to new sites in the genome by misdirected gene jumping could then turn them on inappropriately, resulting in unregulated growth. In other instances, the genes seem to be activated without any change in location by a failure of normal regulatory mechanisms. Many of these oncogenes appear to be transcribed in tumor cells, leading to the production of oncoproteins that may themselves play a critical role in the production of cancer. The important point is that the viral introduction of new oncogenes into the human genome is not required for carcinogenesis. The simple rearrangement or activation of oncogenes, as for example might accompany the exposure of cells to a cancer-causing chemical, appears to be sufficient. It may be that the myriad of chemical and physical agents and factors that have over the years been found to be carcinogenic work in large part by causing the abnormal rearrangement of normal growth genes. This possible common path leading to cancer offers the possibility that a common drug or intervention could block the carcinogenic effects of multiple offending agents.

Additional support for the concept that oncogenes represent the simple misplacement or reactivation of normal genes has come from recent studies that combine the techniques of Bioburst science with those of the computer revolution. PDGF as previously noted is a well-established growth factor responsible in large part for wound healing and also for atherogenesis. Recently the gene coating for this protein was cloned and sequenced. Like all such sequences, it found its way into computers around the world for possible use as a reference standard against which various other genes could be compared. Dr. Russell Doolittle of the University of California at San Diego entered the PDGF nucleotide sequence into his computer bank of gene sequences and asked the machine to report to him the identity of any other genes with similar sequences (homology) that were found in the computer's memory bank. The machine turned up a gene that was remarkably homologous to that coding for PDGF, and that gene, it turned out, coded for the simian sarcoma virus oncogene. The same correlation was found by Dr. Michael Waterfield of Great Britain. This discovery suggests that some oncogenes may produce cancer by overproducing growth factors, which possibly then lead to overstimulation of the division process with cancer being the end result. Growth factors may play an important role in cancer after all.

It is too early to tell if other oncoproteins have growth-promoting activity, but the lead is very hot and offers, should a role for growth factors in oncogenesis be confirmed, many potential therapeutic and prophylactic possibilities. One of the more obvious is that monoclonal antibodies to oncogene growth factors or their cellular receptors could conceivably either prevent or stop the growth of some cancers. While the clinical application of this discovery to patients, if indeed the hypothesis is correct, may be years away, the possible existence of so direct a mode of attack on cancer progress is nothing less than staggering.[5]

Perhaps more important, however, this recent report of homology between an oncogene product and a growth factor serves to confirm the molecular biological commonality of diseases so diverse as heart attack and cancer. At the same time, it again demonstrates the conceptual and technical power of molecular biology, for this science has provided both the insight that these diverse diseases are intimately related and the technical means for exploiting and applying these observations.

The existence and true nature of oncogenes leads to yet another interesting observation. If oncogenes are normal human and animal genes, it is likely that the oncogenic viruses acquired these genes during their multiple entries and exits from mammalian genomes.

Viral oncogenes are probably not viral at all! Actually, the story may go even further, with virus potentially serving as a source of natural genetic exchange between higher forms of life, such as mammals, and less-complex forms, such as bacteria. The classical description of evolution described the gradual development of more and more complex genetic forms—it never really considered that the higher forms could feed genetic information back *down* the evolutionary ladder. And yet this may occur in nature, thanks to the retroviruses.

Of even more interest is the prospect of deliberately using replicative-defective retroviruses to shuttle genetic materials into the cells of higher organisms. This would potentially be of great value to genetic engineering, for although methods exist for the introduction of genetic material into mammalian cells, they are relatively inefficient. For example, the introduction of genes into the cells of higher organisms can be accomplished *in vitro* by arranging culture conditions so as to enhance the internalization of DNA simply added to the cells in culture. Once in the cells, added DNA is processed by housekeeping enzymes and eventually is parked somewhere in the genome. Thus, the introduction of new genetic material into mammalian cells can be accomplished even without retroviruses. Moreover, newer technologies of cotransformation involve the transfection of mammalian cells not only with the gene to be introduced but also with helper genes to provide the transformed cells with a competitive edge for survival in the body. This in itself is a startling idea.[6]

However, retroviruses potentially offer the opportunity to add DNA more efficiently compared to other methods. In the near term, retroviruses will likely be used therapeutically to introduce genes into cells isolated from patients. For example, cells obtained from patients suffering from inherited enzyme deficiencies could be treated with retrovirus containing the gene for the missing enzyme. The treated cells could then be reinfused into the patient to produce a cure. If replicative-defective virus is used, there would be at least initially no risk of establishing an infection capable of spreading genes to other humans. Of course, it must be recalled that some replicative-defective retroviruses can replicate with the assistance of so-called "helper viruses," which can infect a cell harboring replicative-defective retrovirus and lend to the virus the replicative machinery in which it is deficient. The retrovirus then can multiply with subsequent spread to other cells, perhaps even to cells in other organisms. Strategies for the use of viral vectors in human gene therapy must take this and many other potential pitfalls into consideration if these techniques are to be effectively and safely used. Nonetheless, there is every reason to believe that these methods can be employed. Moreover, recent studies suggest that it may be possible to utilize techniques of homologous recombination in gene therapy. It appears that, in yeast, introduced genes neatly displace their endogenous counterparts from the genome by a mechanism that is not yet understood. If this process can be clarified, it is not impossible that it could be put to use for genetic therapy and perhaps lead to the construction of entirely new vectors possessing insertional specificity and little or no capacity for rescue. Although the potential advantages of using retroviruses as shuttles for human DNA are speculative, it is hoped that these possibilities will at least show the virus in a new light—as a potential pharmaceutical. It will be recalled that this class of viruses, for all its malignancy in terms of cancer induction in animals and man, also gave us reverse transcriptase, and clarified the nature of oncogenes and potentially of human cancer; it is the retroviruses that in future may serve as pharmaceutical vectors for gene therapy. Retroviruses truly lie at the heart of Bioburst. This is but one of the small ironies in the biological revolution through which we are living.

The practical applicability of much of what has been discussed in this chapter may seem remote. But this perception derives more from the short time during which modern biotechnology has been available than from any failing on the part of that technology. To the reader who asks what good knowledge of oncogenes does him today, as opposed to five or ten years from now—admittedly a particularly shortsighted point of view—it can be answered that oncogene technology is helping to stem what has been described as the worst epidemic in modern American history—AIDS.

This story begins with a particularly severe form of leukemia that in recent years has been identified in parts of Japan. This disease is unusual from several points of view. First, the leukemic cells are T lymphocytes (T cells), not the B lymphocytes (B cells) more commonly seen in leukemias. B cells are those lymphocytes that make antibodies and, as has been noted, oncogenes can cause cancerous changes to develop in normal B cells. T cells are regulators of the immune system, secreting factors that orchestrate all aspects of the immune response to an antigenic challenge. T cells are the prime movers in "cell-mediated immunity," that form of attack on foreign antigens that does not require the presence of circulating antibodies specific to the antigen; rather, specific sets of T cells seem to react to specific antigens based on recognition receptors on the T cells themselves. T-cell physiology is complex and only now is beginning to be elucidated. But there is no doubt that proper T-cell function is required for the normal functioning of the immune system in the face of foreign antigens—and possibly, according to some authors, in the face of spontaneously developed cancers as well. This latter hypothesis—the so-called *immune surveillance theory*—holds that from time to time small spontaneous cancers develop in animals and man, only to be snuffed out by the T-cell-directed cell-mediated immune system. Only when this system for some reason is unable to destroy a small focus of cancer does clinical disease develop. Although not entirely proven, this idea derives considerable support from medical research. The possibility of immune surveillance cannot help but add to one's appreciation of T lymphocytes. It is these cells that become cancerous in human T-cell leukemia.[7]

Dr. Robert Gallo of the National Institutes of Health has for some time contended that human T-cell leukemia of the Japanese sort is virally transmitted. Indeed, he has isolated a virus that homes in on T cells and renders them cancerous. Antiserum has been produced that recognizes this human T-cell leukemia virus (HTLV-I) and can be used to detect the presence of certain viral antigens in tissues and blood. Additional tests have been developed to detect the presence of antibodies to HTLV in patients' blood, the idea being that a person exposed to HTLV will make antibodies against it. Even more impressive is the development, by the technique of nick translation of the viral DNA, of probes that can be used for hybridization study of cells and tissues. These procedures permit the identification of viral DNA in infected cells. With these tools, principally the first two, it was shown that HTLV exposure is common in parts of Japan; the evidence also demonstrated that although the virus is not highly contagious, the infection certainly can spread *in utero* from mother to child, and other modes of transmission (venereal, for example) appear to be possible. It is important to note that not all infected patients develop leukemia, but a small fraction of such people are unlucky enough to integrate the HTLV viral material into the human genome in a way that induces leukemic transformation.

The evidence is virtually inescapable that HTLV is an infectious virus that can produce leukemia. The world community must note that this virus is not limited to Japan. HTLV or a closely related virus has been found in the Caribbean and is now making its appearance in the southeastern United States. HTLV is potentially a health problem of considerable importance—and not only because of its relationship to leukemia.

Before pressing on with the HTLV story, we should note explicitly what the reader may already suspect. HTLV is a retrovirus. It appears in some ways to be a human analog of feline leukemia virus—an agent well studied by biologists in recent years. Thus, the biology of the retroviruses that was discussed earlier can be seen to assume dramatic importance for everyday life. To be sure, the association of viral infection with cancer in man has been suggested before. Epstein-Barr (EB) virus, the agent responsible for causing mononucleosis, has been associated with a malignant disease called Burkett's lymphoma, as well as with head and neck cancers. True, EB-virus infection seems to be much more frequently associated with lymphoma in Africa than in the United States, suggesting that some second event or condition is required for malignant transformation. Hepatitis B virus (serum hepatitis), although usually a benign disease, sometimes induces a chronic carrier state in its victims. In this case it confers a substantially increased risk of liver cancer. Herpes virus, a group of viruses causing the common cold sore as well as genital herpes, appears to increase cancer risk in its victim. Women harboring genital herpes seem to be at increased risk for cervical cancer. Other examples could be given. But in all these cases—

Epstein-Barr virus, herpes virus, and hepatitis B virus—the agent involved is a DNA virus and the link to cancer is tenuous, involving the passage of long periods of time and possibly encounters with other insults, such as an additional encounter with a retrovirus that facilitates the induction of cancer. HTLV, on the other hand, is a retrovirus. It is a member of a class of RNA viruses known to cause cancer relatively quickly and in a fairly direct and infectious way. The discovery of HTLV should lead everyone to sleep just a little less well.[8]

But there may be still more to this story. In 1981, a New York dermatologist, Dr. Friedman-Kien, was surprised to see several cases of a rare but lethal skin cancer, Kaposi's sarcoma, in young homosexual men. He alerted colleagues in San Francisco and other cities, and similar cases were soon discovered. The rest is history. Studies of these and a growing number of comparable patients led to the discovery not only of cancers—most commonly Kaposi's sarcoma—in these patients, but also of a host of infectious illnesses none of which is common in most people. Yet in these homosexual men such illnesses occurred frequently and often were lethal. Bizarre or "opportunistic" infections are not new to medicine and have been known to occur in patients treated with immunosuppressants, a powerful class of drugs that are used to treat severe autoimmune disease and transplant rejections. But the opportunistic infections afflicting the homosexuals were clearly even more severe and numerous than was usual in patients treated with immunosuppressants, so it was argued that these men were suffering from a severe acquired form of immunodeficiency. Detailed studies of the immune systems of these patients revealed marked abnormalities of T lymphocytes, including a reduction in the so-called "helper" class of T cell. Thus, the syndrome can be explained as follows: loss of normal T-cell function (cause unknown) leads to loss of immune competence, which in turn leads to opportunistic infection (because the body has lost its usual immune mechanisms, and organisms normally controlled by the body run rampant) or to cancer (either because of the loss of immunosurveillance, infection or co-infection with a cancer-causing agent, or some other indirect mechanism associated with the etiology of the T-cell dysfunction). Given this scenario, it is no wonder that the disease, acquired immunodeficiency syndrome, or AIDS, is lethal.[9]

In the early days of the epidemic it was suspected that drugs or other chemical agents could be the cause of the disorder. But as time passed and the disease spread, this hypothesis became untenable. It was found that blood transfusions from AIDS victims could transmit the disease, as could clotting factors that were used to infuse hemophiliacs. Thus hemophiliacs dependent on receiving large quantities of blood products to prevent bleeding became the second major group at risk for the disease. European nations suddenly began to balk at accepting blood products derived from pools of American donors, and many patients in hospitals across this country elected to donate blood in advance of planned surgery in order to avoid the risk of being infused with potentially contaminated blood. The idea that the nation's blood supply was poisoned, however inaccurate from a statistical point of view, resulted in a reduction of blood donation (because of the misconception that donating blood could result in disease) and at the same time produced considerable consternation among patients and surgeons who feared that elective surgery would be curtailed because of diminished blood supplies.

Additional groups demonstrated an abnormal prevalence of the disease. The first were drug addicts, who commonly share unsterilized needles and were found to develop the disease at a significant rate. AIDS also appeared in Haitian immigrants who were not drug addicts, hemophiliacs, or homosexuals. These findings led to the supposition that AIDS was caused by a virus, possibly endemic in Africa, which had been acquired by visiting homosexuals with subsequent transmission by either sexual contact or blood products to other groups. Because both the putative AIDS virus and HTLV seem to attack specifically T lymphocytes and because both are known to exist in the Caribbean, evidence of HTLV infection was sought in AIDS victims. Serological (antibody) testing, gene hybridization studies, and direct viral culture confirmed the presence of HTLV or an HTLV-like virus in some, but certainly not all, AIDS victims.

Although HTLV could have been just another opportunistic infection in these patients, this seemed unlikely since HTLV is generally uncommon in non-AIDS-infected patients in the United States. On the other hand, the absence of evidence of HTLV infection in all AIDS victims can be explained in several ways. First, the tests directed at HTLV virus may not have been entirely reliable because the AIDS virus may be similar to HTLV but distinct, immunologically and genetically, from it. After all, HTLV causes leukemia, whereas the AIDS agent causes immunodeficiency sometimes coupled with Kaposi's sarcoma. The diseases are not identical, and so the available probes (both immunologic and genetic) may not have been as sensitive as one would like. Another possible explanation stems from the fact that the HTLV genomes were demonstrated by hybridization techniques in cells of one AIDS patient early in the course of his disease but later apparently disappeared. The simplest interpretation of this finding is that the T-cell targets of AIDS are in time killed by the HTLV virus, leading to a loss of T cells—and coincidentally therefore to a loss of viral genome from the body. Other explanations can be offered as well, but it is sufficient to note that the absence of evidence of HTLV infection in all AIDS patients was quite properly not taken as conclusive evidence against HTLV participation in AIDS. In retrospect, the HTLV infection detected in AIDS patients may have resulted from opportunistic infection by the virus but the work went on.

Then scientists in France, headed by Dr. Luc Montagnier, detected evidence of infection by a retrovirus in some patients with AIDS and argued that this relative of HTLV—called *lymphadenopathy-associated virus,* or LAV—rather than classical HTLV is actually causally associated with AIDS. Finally, the Gallo group detected a new HTLV-related virus—so-called HTLV-III—in AIDS patients. Evidence of infection by LAV was found in about 90 percent of AIDS patients by the French, and evidence of infection by HTLV-III was found in about 90 percent of patients by the Americans. This leads naturally to the conclusion that HTLV-III and LAV are either the same or very closely related, and that this virus is the cause of AIDS. (Because HTLV-III causes AIDS rather than leukemia, the HTLV designation has now come to stand for "human T-cell *lymphotrophic* virus," meaning the viruses of this group home in on T-cells to cause either leukemia or AIDS.) While not every person infected with the virus develops full-blown AIDS, many do. Indeed, hybridization analysis of the lymphocytes of AIDS patients suggests that coinfection with hepatitis B virus may be one factor that predisposes the patient to the development of clinical disease. The story is still unfolding, but it is clear that LAV/HTLV-III lies at its heart.

Another interesting point related to a possible role for HTLV in AIDS has recently appeared in the scientific literature. Studies have revealed that feline leukemia virus, the animal analog of HTLV, which kills many cats through the induction of leukemia, kills even more by inducing a syndrome of immunodeficiency. AIDS may thus be a retrovirus-associated phenomenon that afflicts many species—including man. This observation helps make palatable to the intellect the fact that cancer-causing viruses are afoot in the world and that they can cause epidemics of acquired immunodeficiency—among perhaps other disorders.

Still more recently a possible additional benefit of science's newfound knowledge of HTLV has emerged. Suramin, a drug used in the therapy of the parasitic disease Rhodesian trypanosomiasis, has been reported to inhibit the reverse transcriptase of many animal retroviruses. With the discovery that the retrovirus HTLV-III plays an etiologic role in AIDS, this observation assumed new significance. The Gallo group has demonstrated that suramin can, in the test tube, block the infectivity of HTLV-III, raising the possibility that this drug could arrest the development of AIDS in patients. While this possibility remains conjectural, it is a source of real hope for progress against AIDS and possibly other retrovirus-mediated disorders.[10] Because HTLV-III can infect cells other than T lymphocytes, including for example brain cells, any drug used for therapy (as opposed to the prevention of infection) must pass the so-called blood-brain barrier and enter neural tissue. One such drug capable of inhibiting reverse transcriptase and entering neural tissues is azidothymidine, which is currently undergoing clinical trials; other similar drugs are under study in the United States, France, and elsewhere. There is hope.

It is too early to predict the final chapter in the AIDS story. The full magnitude of the epidemic cannot yet be predicted, and only recently could the causative agent be identified with reasonable certainty. It is unclear if vaccines against HTLV will be made and if they will prove to be protective against T-cell leukemia and AIDS. What this episode does tell us clearly and forcefully is that molecular biology, including recombinant DNA technology, is not only a promise for the future but also a reality for today. It is impressive that cDNA hybridization techniques will one day be used on clinical biopsy material so that oncogenes can be detected and their genomic locations determined, thereby permitting a diagnosis of malignancy or benignity in cases in which classical pathological findings are equivocal. Even more impressive is the fact that these techniques are being used today to hunt down the killer agent responsible for AIDS.[11]

A similar argument can be made for monoclonal antibodies. As discussed earlier, these ultraspecific antibodies are now being used in many innovative ways to diagnose and treat disease. Monoclonal antibodies are widely used in radioimmunoassay for the diagnosis of disease and are beginning to find a therapeutic role in transplantation medicine and the treatment of those diseases characterized by overactivity of the immune system. If a transplanted organ is undergoing rejection because of proliferation of one class of T cells, the infusion of a monoclonal antibody directed at those cells can sometimes lessen their numbers and mitigate the rejection episode. Advances are also being made in cancer therapy. In those (possibly few) cases in which tumor-specific antigens can be identified, these agents may well be curative, alone or coupled with cell toxins or radioactive isotopes. In some cases in which absolute tumor specificity cannot be achieved, relative specificity may do almost as well. In either case, the hidden locations to which tumors spread can be identified using radioactive monoclonal antibodies and standard scanning techniques. These specific antibodies are now being used to remove leukemia cells from samples of bone marrow taken from patients suffering from leukemia. Once these samples are cleansed with monoclonal antibodies, all the marrow remaining in the patient (including all leukemic cells) can be destroyed with drugs and high-dose x-rays. The patient's cleansed marrow sample can then be reinfused to repopulate his bone marrow with normal cells. This technique, which is currently undergoing development, could have wide application in the treatment of blood cancers.

Monoclonal antibody technology is now undergoing a phase of development prior to widespread use, just as a prototype aircraft is test flown and modified prior to being mass-produced for routine service. Although we must await the results of this process of development, there is every reason to be optimistic about the outcome.[12]

Bioburst has opened new fronts in man's battle with cardiovascular disease and stroke. It has discovered the common denominator that unites diseases as diverse as heart attack and cancer. Because of recently acquired knowledge, the prospect for significant advances against these diseases over the next decade is very bright. There is even more reason for rejoicing over the scientific breakthrough that made this optimism possible and promises ever-greater benefits in years to come.[13]

CHAPTER VIII

Bioburst and
the Germ Line

Our discussion of genetic engineering to this point has centered on somatic cell transformation—that is, the analysis has always involved the genetic interaction between body cells (*soma*) and various pharmacologic agents. Genetic manipulation of germ line cells (the ova or sperm from which future generations arise) has not been touched on. This choice of soma over germ line derived from a desire to avoid temporarily the controversies surrounding the appropriateness, and indeed the morality, of germ line (as opposed to somatic) interventions. Religious leaders have issued resolutions suggesting that any interference with the germ line is immoral, and some observers of the cultural scene, such as Jeremy Rifkin, the author of *Algeny,* have taken a similar tack. The basic argument is that any attempt to alter the human germ line is ill advised, at least in part because this process necessarily leads to loss of genetic diversity—the stuff that biologic progress is made of. Opponents fear that such interference will necessarily lead both to large-scale tampering with the human gene pool for frivolous purposes and, most frighteningly, to a resurrection of eugenics, and that the resulting loss of genetic diversity could then have tragic moral and biological consequences for the species in the not-too-distant future. And there is also the contention that germ line engineering profanes the very essence of man's nature and humanity. Therefore, opponents argue, children born with Tay-Sachs disease, Down's syndrome, or a host of other disorders must be permitted to die without benefit of gene therapy.[1]

This argument deserves detailed study since it potentially sentences thousands of children to death in each generation. Before beginning this analysis, it is worth investigating the ways by which germ line diversity is currently being altered by man-made forces. But when one attempts to address this issue, problems are immediately encountered. For one thing, it is difficult to define genetic diversity adequately. Leaving that aside, what is clear is that many human activities, ranging from the simple preference of many tall people for tall mates to the extreme of female infanticide as allegedly practiced even today in Asia, tend to alter the distribution, and indeed the frequency, of specific genes in the population.

Taking considerable liberties with modern terminology, we can arbitrarily divide all processes that alter gene frequency into those that are subtractive and those that are additive in their initial effect. By a *subtractive* force, I mean an intervention that has as its initial effect the elimination of genes from all or part of the total gene pool. Because genes appear to be dynamic, as we have seen, subtractive effects in one generation may lead to a selective advantage for remaining genes, which could then evolve by recombination and other mechanisms in ways that might otherwise have been denied to them. Thus new genes may in subsequent generations be added to the gene pool. It can be appreciated, then, that an initial subtractive effect may, at least in part, be associated with an additive effect at a later time. Similarly, if genes are initially added to the gene pool, either by mutation, retrovirus infection, or genetic engineering, the effect will be termed *additive*. However, these effects may lead to the extinction of other genes by selective pressures in subsequent generations. Therefore, there may be few purely subtractive or purely additive gene manipulations.

In a practical sense, however, subtractive effects can have a devastating effect on a species, particularly if the subtractive forces produce a dramatic reduction in the species' gene pool. The effect of the magnitude of a subtractive force on subsequent genetic diversity can be illustrated by considering an example. If all tigers, save those few on a remote hilltop, drowned in a great flood, then all subsequent tigers would for generations be closely linked genetically to those few hilltop cats, and a major loss of genetic diversity would have occurred. Some such event must have transpired in the development of the modern cheetah, because all living members of this species display remarkable genetic similarity. It appears that the cheetah has, as a species, passed through an evolutionary "bottleneck" such that only a few animals became the progenitors of the modern species. And the cheetah has suffered severely for this. The modern cheetah, for all its beauty, is a frail animal with a poor ability to fend off illness. Indeed, it was the manifest lack of vitality in the cheetah that led zoologists to undertake the genetic studies that detected the relative homogeneity of the animal's gene pool. In the case of the cheetah, the loss of genetic diversity has led to loss of vitality and has lessened the ability of the species to survive if a change in environment should lead to the elimination of the comfortable ecological niche the cheetah now occupies. Loss of genetic diversity therefore appears to lead to loss of vitality and adaptability, so major subtractive genetic changes that occur in the face of a developmental bottleneck or the like can be severely damaging to a species.

In a real sense it is the magnitude ("dose") of the subtractive change that is harmful in these cases. To paraphrase this point, the loss of all or a large fraction of the genes for a given trait in a given species is qualitatively different from the loss of some of the genes for that trait. In this regard it should again be noted that a wide variety of human activities produce small alterations in human gene frequency without any apparent harm to the species. Even the somatic treatment of genetically influenced diseases, such as diabetes, alters gene frequency in subsequent generations. Insulin and other therapies prolong the lives of diabetics so that they can propagate their genes. A similar analysis can be applied to the use of glasses for myopia, which prevent patients from being hit by cars at an early age, as well as to drugs for hypertension and the hyperlipidemias that prolong reproductive life. Much of modern medicine has genetic consequences, and yet no one objects—in large part because of the small "dose" of genetic change involved. If, for example, the diagnostic power of molecular biology were to be aggressively used for the purpose of aborting embryos with less than desirable characteristics, the subtractive dose to the genome of any such activity would have to be considered in determining the potential biological effects of the procedure on our species. It is to be hoped that the use of common sense will restrict these techniques to the elimination of disease, but in any case the dose of the induced genetic change would seem to be a factor of critical importance in determining the net effects of the intervention.[2]

It could be argued that the above examples produce only variations in the prevalence of various genes in the population and that genetic engineering results in quite different consequences. Both sides of this argument require analysis. First, it is not at all clear theoretically that the examples of genetic alteration just considered change *only* gene prevalence. This view assumes that genes are truly atomic—indivisible and indestructible—and yet modern genetics is beginning to teach us that genes are dynamic, capable even of leaping large distances. They may not be so indivisible as once thought. Even if genes are more or less "atomic," changes in gene prevalence if sufficiently large can lead to gene frequencies inadequate for the survival of some traits while others not only persist but gradually evolve in new directions. Changing gene frequencies is a potent mechanism for producing biological change. In fact, "change in gene frequency" has been called the "official" definition of evolution.[3]

The second arm of the argument is that genetic engineering promises to do something quite different from what has gone before. To assess this claim, it is necessary to decide what germ line genetic engineering actually will do. The experiment that prompted much of the recent discussion about this subject involved the introduction of extra growth-hormone genes into fertilized mouse eggs. The ova were then returned to the womb with the eventual production of a giant mouse. The artificially added growth hormone genes functioned to produce growth hormone, thereby resulting in an excess of this material and a state of gigantism. This achievement by Richard Palmiter, Ralph Brinster, and colleagues sent shock waves through the scientific and philosophical communities. It is easy to extrapolate these results to assert that in the not-so-distant future it may be possible to insert the gene for normal hemoglobin into patients lacking it (and therefore condemned to die at an early age from thalassemia), resulting in normalization of life expectancy. If this genomic transformation is produced in the ovum or early embryo (as opposed to cells isolated from an adult), or if in introducing these genes into an adult they are simultaneously incorporated into both body and germ cells, then this gene change may be passed on indefinitely through the germ line into future generations. Moreover, all genes associated with thalassemia would similarly be passed on at a rate that would be denied them in the natural state. These two points, the passing on of new genes in the germ line and a change in the rate of associated gene transmission, form the crux of the issue. Germ line genetic engineering is forever.[4]

How can saving children suffering from Tay-Sachs or other genetic diseases through gene therapy be harmful? For one thing, the gene therapy itself might be dangerous in some unanticipated way. This danger could arise from the mechanism by which the new DNA is inserted, or the therapy could unmask some other deleterious gene associated with Tay-Sachs disease but masked in the general population because of the lethality of the disease (Tay-Sachs is used here solely as an example; there is unlikely to be any such deleterious genetic cargo associated with it). In either case, if an untoward lethal consequence of gene therapy occurs before reproduction, then the therapy will have only prolonged the patient's life, not cured him. If the untoward effect occurs after reproduction, then the next generation will feel the consequences, and the deleterious gene, all other things being equal, will spread into the general population. To eliminate this possibility, the recipients of germ line gene therapy could be sterilized or their reproductive activity could be carefully monitored and possibly limited, since even these Draconian measures would make more sense than denying therapy to those who need it. However, it is unlikely that these policies will be required for several reasons.

First, a hypothetical deleterious gene is unlikely to spread very rapidly into the general gene pool. Social pressures alone would tend to reduce the number of offspring produced by patients who have undergone germ line therapy, and this would slow the gene's spread. More important, the number of people treated is likely to be so small compared to the total number of fertile people in the world community that it would take many generations for the genes of treated individuals to make significant inroads into the general pool—by which time the safety and stability of those genes would be proven. In the vast majority of circumstances in which genetic engineering is contemplated today, somatic therapy of the child or even the fetus would be used, so the number of people—if any— who will require germ line therapy will be extremely small. Finally, it is likely that, in the decades before this kind of therapy becomes a feasible procedure, techniques will be perfected that permit reliable assessment of the risks involved.

It may be, for the purpose of argument, that genetic therapy by the introduction of new genetic material increases the risk of subsequent cancer. At the same time, hybridization techniques may be able to determine which families, by virtue of retrovirus infection eons ago, are naturally at a similarly increased genetic risk for cancer. A test that uncovers an increased risk for cancer would provide a social disadvantage for patients with either naturally or artificially increased cancer risk. The point is that any increased risk associated with gene therapy is likely to be quantifiable. Doctors may one day counsel young people about to get married that the previous genetic manipulation in their families is "nothing worse than having had mononucleosis." Society in the form of its individual members will then be free to deal with these risks directly in choosing mates.

To argue that the genetic testing that provides this information is as harmful and disruptive to the normal exchange of genetic material as the idea of gene therapy itself is not compelling. Already cancer risk, as well as the risk of potential heart disease and diabetes, can be to some degree predicted from family and personal history coupled with routine blood tests and the like. To decide that genetic screening is intrinsically immoral requires that much of the risk-factor assessment carried out by modern medicine be deemed immoral, at least to the extent that it can potentially be used for sexual selection. However, most people would probably agree that neither risk-factor determinations, future genetic screening tests, nor even germ line therapy itself are intrinsically immoral. And although germ line genetic engineering is dramatically different from current practices of gene pool alteration (in that in theory *arbitrary* changes in genes can be *rapidly* produced), it is not morally different.

To be sure, the difference between germ line genetic engineering and everything that has gone before is significant. As long as the numbers of people so treated are small, there will be little risk of genetic catastrophe, and the benefits to patients who otherwise would have died lingering deaths will be enormous. If the number of people treated with germ line therapy increases, however, the risks become larger. The standard argument is that the trivialization of gene therapy will lead to its use on a larger scale for such things as assuring that offspring have a "desirable" complexion or hair color or for preventing nearsightedness and astigmatism. The commonly expressed fear is that this kind of cosmetic genetic engineering could lead to the loss of, say, dark-skinned myopic people, and could in turn have not only dire social consequences but also unfortunate consequences for the survival of the species (as, for example, if the loss of the ozone layer were to result in increased death rates in a uniformly fair-skinned population). However, this argument is not terribly persuasive. Germ line therapy is clearly a major form of therapy and one that is unlikely to be trivialized either by the common man or by government. It is hard enough to get people to take medicines for major diseases such as high blood pressure and diabetes, much less to get them to alter their germ cells to eliminate myopia in their children. And then again, not all people value fair skin. The combination of a pluralistic society, traditional regulation of therapeutic practices, and common sense will prevent the trivialization of the technology.

A more difficult situation arises when one considers a nontrivial and potentially widespread application of germ line therapy. Suppose, for example, that a minor gene deletion would result in arterial smooth muscle cells with reduced receptor affinities for platelet-derived growth factors. Assume that this change results in normal vascular function but reduces the risk of vascular disease (heart attack and stroke) by 99 percent. It could be argued that virtually everyone would want such therapy and that therefore the risk of any deleterious gene modifications would be unacceptably greater. However, several points can be made regarding this and similar circumstances.

First, in this and all other cases of proposed widespread genetic therapy, a somatic therapy approach would almost certainly be chosen over a germ line approach. That is, a virus or other vector that is trophic (read "specific") for arterial smooth muscle cells would preferentially be employed to alter the receptor genes in the adult or child, and the germ line itself would not be touched. Only in certain relatively rare cases of genetic defects that are deleterious *in utero* would germ line therapy be required, and even in most of these cases targeted somatic therapy of the fetus would likely be used instead of germ line therapy. In the case of some diseases, such as perhaps diabetes, germ line therapy might be more efficient and more easily achieved than somatic therapy, but it is unlikely that it would be required. We can anticipate that widely applied genetic therapy would be of the somatic rather than of the germ line sort and that even selectively applied therapy, such as the treatment of thalassemia, would likely be of the somatic type. In fact, although somatic cell genetic engineering may soon be a reality, human germ line therapy of any sort and on any scale is unlikely to be achieved for many decades. So the entire debate on germ line engineering may be moot for the foreseeable future. In addition, it is likely that many people would reject any proposed germ line therapy as an unnecessary risk, thereby providing a large pool of "untouched" genes. There is a bit of therapeutic nihilist in almost everyone, so it would be reasonable to anticipate that a sizable segment of the population would reject any and all gene therapy and would probably work against germ line engineering in all but the most extreme circumstances. Thus common sense, therapeutic nihilism, and the use of somatic gene therapy will make widespread germ line engineering unlikely in the foreseeable future.

But finally we must turn to the extreme case. In the extraordinarily unlikely event that at some time in the (distant) future (1) a dramatically beneficial therapy can *only* be administered in a way that requires genetic alteration of the germ line and (2) a large segment of population wishes to undertake the treatment, is it so unreasonable to consider offering this therapy after allowing for a period of thorough education in the potential risks of the specific gene therapy involved? Society will have to answer this question based on the philosophical, moral, medical, and scientific aspects of the spe-

110

cific proposal. Even in this extreme case of widespread germ line engineering, an absolute *a priori* prohibition (as opposed to a strong relative contra-indication) against germ line therapies does not appear wise. Much could depend, for example, on (1) the nature and risk of the proposed engineering when compared to natural sources of germ line genetic variation, and (2) the risk to the population— and more particularly to the germ line—that the therapy is designed to offset. Although it is unlikely that widespread germ line engineering will *ever* be required or employed, it nonetheless does not appear necessary, or even wise, absolutely to preclude it in advance. We and our germ line may some day need it. In the remote event that the above criteria are met, the therapy is undertaken, and a genetic accident occurs, there would be ways to counteract any ill effect. In a "worse case" scenario, stored sperm and ova could probably be used to reconstitute the "original" unengineered human genome. More likely, genetic engineering itself could be directly employed to excise whatever gene lesions had been produced.

In this regard, we must recall that fertilized ova of both animals and humans have already been stored frozen for long periods of time. The New York *Times* of June 26, 1983, reported the birth of a filly derived from a frozen embryo. After remaining in deep freeze for two months, the fertilized ovum was thawed and inserted into the womb of a mare, and the filly was the result. In addition to being a boon to the breeders of horses, since a great stallion can now be used to fertilize multiple eggs with the resulting fertilized ova frozen in large numbers for future use, this technology also offers the possibility that people could conceive their children in one decade and bear and raise them in another. Already many children have been born worldwide to otherwise infertile couples employing *in vitro* fertilization and in some cases the use of frozen embryos.

This new technology has already raised a spate of social questions, the most thorny of which derive from what rights, if any, society is to bestow on frozen embryos that for some reason are not implanted into a nurturing womb but are rather left in the limbo of a laboratory freezer. The social consequences of this embryo freezing are hard to fathom, and although it can reasonably be hoped that no such activity is undertaken on a wide scale, it should be pointed out that this technology in and of itself is something of a genetic safety net. At the same time, of course, it is a form of genetic manipulation or "engineering," since there must presumably be some theoretical chance that the genome of the stored ovum will change with time. Parenthetically, it will be interesting to note if normal people will clamor to take advantage of the convenience of this kind of genetic engineering. I think not, just as I doubt that they will demand the trivialization of any form of genetic engineering.[5]

It is also interesting to note that the storage of fertilized ova could be of value to patients with neoplastic diseases. For example, if a woman were to have a serious medical illness or cancer, she might elect to mate with her husband, have the fertilized ovum stored, and then permit her disease to be treated aggressively, even with potentially mutagenic therapies. Having survived her illness, her embryo could then be replanted in her uterus and a child born. There is nothing in this practice intrinsically different from storing the sperm of a man about to undergo chemotherapy and using it in subsequent years to fertilize his wife—a practice now commonly undertaken. Are these immoral acts?

Returning to the argument, the technology that permits the freezing of sperm, ova, and embryos provides society with the capacity to store what essentially amounts to the seeds of the human race. This then becomes a safety net in the event of a major genetic accident as it would in the case of widespread natural plague. Of course, if this should ever be necessary, new moral and technical questions will immediately arise. Should these cellular entities at some point be permitted to resume their development—perhaps to leave in their place some of their genetically unengineered offspring—or should they simply be viewed as seeds in storage? But again, this entire line of argument is extremely hypothetical, as is for that matter much of the discussion in this chapter, because human germ line engineering of any sort is many decades away. The use of somatic therapies coupled with simple common sense will likely render any accident arising out of gene therapy exceedingly improbable—indeed, less probable than a natural infection having genetic consequences of comparable severity. Common sense coupled with the ethos of a pluralistic society will make the rise of a new eugenics based on germ line engineering not only infinitely less likely than many fear, but also less likely than a recapitulation of the eugenic nightmares of the past—all of which occurred without the benefit of modern molecular genetics. In addition, we must recognize that some fears regarding "genetic engineering" and its impact on society are, for the foreseeable future, totally unfounded. For example, it is a common misconception that genetic engineers will come under pressure to make the next generation smarter, better at music, more athletic, or better looking. Genetic engineers are only now beginning to introduce single genes into cells. The capacity to produce characteristics like intelligence or athletic ability, which are under the control of many genes and influenced by the interaction of those genes with environment, does not exist today. There is a world of difference between correcting a single-gene defect in an effort to spare a child's life and attempting to alter multigenic traits. We must not lose sight of this distinction.[6]

In a sense, the present arguments for proscribing germ line therapy represent updates of two arguments faced by Cotton Mather in the eighteenth century: (1) is it intrinsically moral to make oneself ill in the case of variolation or to alter the germ line in the case of bioengineering? and (2) should the "variolated" or "engineered" be isolated from the rest of the population, either physically in the case of variolation or reproductively in the case of bioengineering? As was the case in colonial Boston, a negative answer to the first (moral) question seems to rest on a brand of fatalism that is difficult to reconcile with either natural illness or prophylactic medicine. On the second point, however, even Mather probably had to admit that isolation of variolated subjects should have been recommended for the good of the public and, in particular, of the nonvariolated segment of the population. It must be recalled that if variolation did indeed spread smallpox to the unvariolated, it had the theoretical potential of placing at risk all the people of Boston, who at the time constituted a not-insignificant percentage of the population of what was to become the United States. By analogy, either reproductive isolation or monitoring and control of the spread (dose) of the genetic change produced by germ line genetic engineering ought be instituted. When these techniques and capabilities finally are achieved, they should not be trivialized but rather used sparingly and judiciously, and the degree to which they are practiced should be monitored over time. It is unlikely that widespread germ line engineering will ever be necessary, but if it should be, the community must realize that risk is involved and considerable deliberation and consultation should occur in all segments of society prior to permitting the undertaking. In that event, common sense and molecular biology itself should serve us in good stead. After all, variolation was a success even in the absence of any isolation. In the face of epidemic, even uncontrolled variolation (potentially a massive dose to the population) was preferable to doing nothing; in the absence of epidemic, the risk was deemed too great. So too, in the face of an overwhelming social need for widespread germ line engineering, the procedure may be acceptable to society, particularly given the safety net that molecular biology may soon be able to provide. In the absence of overwhelming need, any such risk is unacceptable.[7]

Although the issues raised by DNA technology will require careful deliberation by the scientific community, the lay community, and government, there is every reason to believe that this technology, like others, can be used safely if common sense prevails. To dismiss it out of hand as immoral is unfair to the ill.

CHAPTER IX

Bioburst and Public Safety

Legitimate questions about safety and ecological impact can, and should, be raised whenever a new technology is introduced. This is all the more true when the technology is biologically directed. Indeed, questions about the safety of cDNA research were raised as early as 1974 and continue to be aired even now. It is likely that these issues will be discussed and reexamined as long as the horizon of biological knowledge continues to expand. Moreover, as the new technology is applied and the public becomes more and more aware of that application, it is likely that these debates will intensify.

The initial concern voiced about recombinant DNA technology in the 1970s centered around the possibility that an *E. coli* bacterium containing a cloned human gene could "escape" from a laboratory and by means of plasmid transfer spread human genes to the wild pool of *E. coli*. It was then feared that a person infected by one of these organisms would not simply develop a bout of diarrhea, as is often the result of *E. coli* infection now, but would also suffer from the effects of the product of the cloned gene. For example, if the gene for a neurotoxin was carried by an *E. coli* ingested by an unfortunate human, enough of the toxin conceivably could be absorbed from the gastrointestinal tract to cause death. Because of concerns like these, a voluntary moratorium on cDNA research was established in 1974 and the subject of cDNA safety was intensively studied. A major conference at Asilomar, California, in 1975 brought leaders of the scientific community together to discuss the full implications of recombinant DNA science. Also, innumerable discus-

115

sions took place in the council chambers of cities where recombinant DNA (rDNA) research was centered.[1]

In spite of this concern, it quickly became apparent that most cDNA research is manifestly safe. The possible exceptions are studies involving known pathogens. These activities continue to be viewed as potentially dangerous, and even the recently relaxed National Institutes of Health (NIH) guidelines continue to require varying levels of biological isolation for those uses of recombinant DNA technology that seem to be associated with some theoretical risk. Although it is virtually impossible for cloned genes to leave *E. coli* and infect human cells, prudence dictates that *E. coli* carrying potentially harmful genes be handled with the same care given to research on other potential pathogens. This is not to say, however, that the risk of studying, say, a pathologic virus in the laboratory is comparable to the risk of studying a cloned human gene, for even were an *E. coli* carrying an oncogene to escape and "infect" wild strain *E. coli* outside the laboratory, there are antibiotics that could be used to stop any impending epidemic in its tracks. *E. coli* is a well-studied organism (even its plasmid-resistance factors are well understood) and one that could probably be contained with available pharmaceutical agents. Moreover, the strains of *E. coli* used in genetic engineering are deliberately chosen for their frailty and inability to compete with wild-type *E. coli* outside the laboratory. This "biological containment" is an important safety factor. In addition, there is no obvious way by which a cloned human gene could make its way from an *E. coli* to a human genome. (One could postulate some form of viral transfer, but any such mechanism could presumably operate to shuttle genes even in the absence of genetic engineering.) Thus, even in the case of oncogene cloning, recombinant procedure can be carried out with considerable safety. In the case of simple hormone gene cloning, the safety factor is still greater because virtually all protein hormones are not absorbable from the gastrointestinal tract, the usual site of *E. coli* infection. Thus, even were a breed of insulin-producing *E. coli* to escape and eventually infect the gastrointestinal tract of a man, virtually none of the produced insulin would be absorbed by the victim. These considerations do not mitigate against care and regulation in this research, but they do bring home the point that an escape of Lassa fever virus from a research laboratory is infinitely more dangerous than an escaped *E. coli* carrying a human gene.

Since the lifting of the voluntary moratorium on recombinant DNA research, an enormous increase in gene cloning has occurred—and in perfect safety. The number of laboratories performing recombinant DNA research as well as those utilizing these techniques for commercial application is growing daily, and no significant accidents or potential accidents have been reported. However, this safety record, and the reassurances given above, hinge on the use of *E. coli,* or agents with comparable safety characteristics, as the cloning vector and on the exercise of barrier precautions in the study of pathogenic vectors. There is no doubt that if analogs of human oncogenes were cloned using a human virus as the cloning vector (an experiment that is highly unlikely, except perhaps in a biological warfare laboratory), and if barrier techniques were to fail, an epidemic of infectious cancer *could* result. This would be an analog of what appears to have occurred naturally in the cases of the lymphoma-causing virus HTLV-I and the AIDS virus HTLV-III. There is a potential danger in this kind of activity, and there is every reason to regulate it, even to restrict aspects of it, in the same way that the study of other potentially dangerous processes is regulated and restricted. However, if precautions are taken, it can reasonably be anticipated that the promise of the new technology will be safely achieved. Oversight and regulation are necessary but need not hinder significantly the conduct of research and the *cautious* application of recombinant DNA technology. Now is the time for science, industry, government, and the citizenry to establish the boards and guidelines that will provide oversight, but not obstruction, as the new science develops.[2]

The analysis just presented suggests that although small-scale accidents can possibly occur in DNA laboratories, as at other sites of research or industrial activity, the possibility of a major accident or disaster is remote if established guidelines regarding vectors and the types of genes cloned are adhered to. But what about the risks operative in other circumstances—such as biological warfare or terrorism? Here it is difficult to analyze risks with much precision.

After all, if a nation were to apply the talents of its best biologists to the deliberate development of harmful biological agents, it is hard to be sanguine about the outcome. In a world in which mycotoxins allegedly are being used in several areas as agents of war, the possible hostile use of newly designed organisms must be taken seriously. These caveats and disclaimers notwithstanding, there may still be certain principles that can be utilized to assess the nature and magnitude of these risks.

First, most bacteriologic or viral epidemics spread relatively slowly. There are exceptions like viral influenza, which can spread across a continent in ten or twenty days, but even this rapid spread takes considerably more time than the thirty or so minutes required for an ICBM to fly from Colorado to Moscow. Even in the case of an influenza epidemic only a minority of the national population actually becomes ill. If a massive lethal epidemic were to spread suddenly across the United States, we might reasonably anticipate that modern molecular biological techniques would relatively quickly determine if a man-made agent was at the root of the problem. It is therefore doubtful that any aggressor nation would risk such an adventure, given the likelihood of nuclear retaliation. It could be argued that an ultrafast viral agent capable of exterminating national populations in a matter of days might not provide the scientific community with adequate time for detailed investigation, much less the design of effective countermeasures, in which case the bioweapon would be an analog of the neutron bomb—killing people but not destroying material. The fact that in the history of mankind no pathogen of such rapidity of action has ever been detected may convince the afflicted nation that biological warfare was afoot, and the anticipated response could be equally vicious biological retaliation, or even nuclear attack. The risks of isolated biowar appear to be unacceptably high for the attacker, in that the prime advantage of biological weapons is anonymity and avoidance of associated responsibility. Modern science could likely assign responsibility relatively quickly. For this reason it is unlikely that these weapons could realistically be employed against a technologically sophisticated society.

There are weaknesses in this argument, of course. It could be argued that, were a rapidly active bacteriological agent loosed on a nation, the responsible governmental officials could not with any certainty know against which adversary to strike. However, this and other counterarguments would not appear to give much security to any potential attacker. Rather, the political context in which such an event may arise, coupled with whatever scientific investigation could be performed prior to the time of latest feasible retaliation, would likely be adequate to clarify substantially the source of the attack. Thus, the use of bioattack as an isolated tool for aggression would appear to have only minimal advantages (those related to possible concealment of the source of the aggression) over the use of nuclear weapons when employed against a technologically sophisticated society.

On the other hand, it could be argued that biological weapons can be effectively used either as part of a larger conventional or nuclear attack or as a component of guerrilla warfare or population suppression in remote parts of the world. In the former case, it is doubtful that these agents would have a major impact in the midst of a modern theater of nuclear war. Their use against populations or guerrillas is perhaps more realistic, as is demonstrated by the alleged yellow rain bioattack on Vietnamese and Afghan resistance elements. But in the absence of total war, the adverse publicity associated with the use of such weapons for guerrilla suppression is likely to be significant, as might be the retaliatory delivery of similar bioweapons by the West (or East, whichever the case may be) to the guerrillas.

For these reasons, it is unlikely that Bioburst will provide decisive weapons in the contest between East and West, although the uncertainties in the analysis render this conclusion relatively soft. It is the best that can be arrived at currently.

The possibility that terrorists might utilize weapons based on recombinant DNA technology is even harder to deal with. Adverse publicity and the need for quick action are not factors that necessarily influence terrorist tactics. It could be argued that "total damage potentially inflicted" is the essence of the terrorist threat, not rapidity of onset or secrecy. Thus biological weapons, including and perhaps especially those developed by rDNA technology, could potentially be employed in terrorist activities. However, the use of engineered life forms in terrorist activities would require either access to ready-made biological weapons or considerable technical skill to develop such weapons. Also the risks to the terrorists themselves would be great—although this factor does not always dissuade such people. It should be noted that the number of people capable of undertaking these rDNA activities is small, a fact that provides some relative measure of security. When compared to conventional agents, DNA technology is likely to add only marginally to the capacity of any scientist inclined to produce mayhem.

Nonetheless, risk exists, and stringent regulation and monitoring of potential pathogens should be undertaken. How best to undertake such a police action is unclear at the moment, but such simple devices as logging personnel trained in these techniques as well as keeping an ongoing list of agents and projects under study in all laboratories would seem reasonable. This information could be kept at the university level and made available to regulatory agencies as required. Although this policy would involve some infringement of the privacy of scientists, it would be a small price to pay for enhancing the security of society. And if such a policing program were properly monitored, it would interfere little with productivity or personal rights. The potential abuse of such an oversight program does exist with the possible pirating of ideas from one laboratory to another, but this could also be contained. If only specific information, such as the training and identity of the responsible investigators and technicians and the pathogens and vectors being employed, were reported and recorded, the policing program would probably achieve its maximal potential for safeguarding society without interfering with proprietary rights or scientific priority.

An additional risk—really a variation of the terrorist theme—that should be considered in this context is the possibility that the biotechnical resources of a nation bent on aggression could be used to support an international terrorist group, thereby shielding the identity of the aggressor. If bioagents developed abroad were made available to a domestic terrorist organization, a more frightful scenario is developed. Because of the possible avoidance of responsibility by the aggressor nation, this course of action might appear appealing to the attacker. But again it is likely (or not sufficiently unlikely to be compelling) that responsibility could be traced to the aggressor state by domestic intelligence agencies. The use of this form of biowar therefore seems too risky to be chosen by a rational aggressor. In any event, the genie is out of the bottle now and there is little in the way of domestic policy that could bear on this issue other than the supervision of potentially dangerous work conducted in domestic laboratories. This area of the misuse of biotechnology now properly belongs to the province of defense planners—just as bacteriologic warfare always has.[3]

An additional and less hypothetical risk posed by rDNA technology is the potentially adverse consequences attendant upon the deliberate introduction of biologically engineered organisms into the environment for agricultural or other purposes. Such organisms could conceivably have significant ecological impact. Already, with the approval of the Recombinant DNA Advisory Committee of the National Institutes of Health, genetically altered bacteria are being prepared to be "released" in order to lessen frost damage to plants. Threatened court action and the recent appreciation of the fact that the release of engineered organisms into the environment likely falls within the purview of the Environmental Protection Agency have served to delay, but only delay, plans for these experiments. Proposals to utilize genetically altered bacteria to detoxify soil or "chew up" oil spills are even now on the drawing boards. Could these well-intentioned microbial releases have unanticipated adverse consequences? Of course! One has only to consider the kudzu growing along a southern highway to see this.[4]

During the Depression, southern farmers planted kudzu, a hearty vine originally imported from Japan in 1876, in drought areas, hoping to protect the soil from erosion. The plant not only did its job but took to the South with a certain glee. It has been spreading ever since, killing other trees, bushes, and hedges and just generally being a nuisance. By analogy, it is certainly not impossible that a released bioengineered organism could in time prove to be a nuisance—or worse. Any release of bioengineered organisms must however slightly or transiently alter local ecology. Whether that alteration will be significant or not depends on several factors. As Dr. Martin Alexander of Cornell University has pointed out, the probability that an engineered organism will do harm is related to the probabilities that the organism will (1) be released into the environment; (2) survive in the environment; (3) multiply in the environment; (4) move to distant sites; (5) transfer genetic information to other species; and (6) act in a harmful manner. Each of these probabilities can in theory be estimated to provide an index of risk in any specific case, and the deliberate genetic engineering of organisms to lower these probabilities could be one way to assure that free-release activities are not injurious to man or the environment. Nonetheless, if a released bioengineered organism possesses the capacity for easy spread or rapid transfer of genetic traits to other organisms, its ecological disruption, however minor, could soon become geographically widespread. Presumably, any such engineered organism would be nonpathogenic, but in selected debilitated hosts, such as patients undergoing immunosuppressive therapy, AIDS victims, and the like, these creatures could act as pathogens. This need not surprise us, since any natural organism can act as a pathogen in an immunocompromised host. This possibility, therefore, should properly not be raised as an argument against the idea of using bioengineered organisms in agriculture. Nonetheless, we can expect that in time some immunocompromised person will die as a result of infection by a bioengineered organism. We must expect to hear in the future of deaths from pneumonia caused by oil-eating bacteria, just as the citizens of colonial Boston had to be prepared to accept some deaths from variolation. Irrespective of any logical cost analysis that might be presented, the benefits of cleaning up oil spills expeditiously and thereby staving off the pollution of the food chain are unlikely to be readily perceived as worth human lives.[5]

Additional problems could arise if released organisms significantly change ecology in unanticipated ways, so that valuable crops are destroyed or rendered less productive. Just as the introduction of genes into a species tends to limit the options for other genes, so too does the introduction of a new species into the ecological environment alter the opportunities for natural organisms. These questions can become quite complex, and accurately predicting ecological changes is a notoriously difficult science.

On the plus side is the fact that bioengineered life forms offer palatable alternatives to chemical insect-control agents and defoliants (after all, bacteria don't cause cancer and some chemical agents currently in use may). Additionally, these new life forms offer the possibility of detoxifying our already polluted environment. They provide enormous possibilities for increasing crop yields and feeding the world's hungry.[6] They possibly can make fertile, and life supporting, arid and sterile regions of the earth, improving thereby the lot of millions. Soon new varieties of corn and tomatoes will be grown, thanks to genetic engineering. The promise of bioengineering is enormous and is not to be lost or fettered unnecessarily.

On the other hand, the lesson of the pesticides has taught us that caution must be exercised in ecological matters, and principles of prudent husbandry must be followed at all times. Now is the time for science, industry, and government to evolve guidelines and oversight procedures to protect the environment while permitting biological engineering to continue its progress efficiently but safely. At all times, it must be clear that the release of engineered organisms is intrinsically more dangerous than run-of-the-mill laboratory recombinant DNA work. Free-release activities must be discussed widely by the public, carefully regulated, and diligently policed. And because life respects no international boundaries, regulation and oversight of free-release experiments should be international in scope. It is to be anticipated that famine-wracked nations of the Third World might be less prepared and less willing to practice strict regulations and oversight than nations in the affluent West. But the West must work closely with these nations as they undertake bioengineered agriculture to assure the safety of all. Biotechnology should only be provided to foreign governments with the understanding that oversight is to accompany it. As Senator Mark Hatfield noted in a recent discussion of genetic engineering and medicine, it is important "to engage in an informal debate with the hoped result that an effective ethic commission, quasi-independent agency, appropriate sanction, or world-wide moral consensus could be developed." At the same time, the senator further expressed his "heartfelt concern . . . that we will not run off half-informed and stir people's fears unduly, impose legislation that will be counterproductive later, or be so apathetic that the technology will once again outstrip the human ability to control it." [7]

In sum, rDNA technology is essentially safe as currently practiced. The use of *E. coli* plasmids for hormone production and similar activities is probably no more risky than routine diagnostic bacteriology. Studies employing potentially harmful genes or vectors that could in theory affect man or livestock are possibly dangerous and should be subjected to strict regulatory control and oversight. Studies or commercial ventures that involve the deliberate release of engineered organisms are also potentially dangerous and should be subject to intense study prior to, and after, being approved. This study should involve ongoing monitoring of the ecological inroads made by the bioengineered organism over time. It would appear reasonable to require that requests from industry or academia to undertake free-release experiments be accompanied by ecological-impact studies, detailed programs for ongoing monitoring, and, in most cases, contingency plans for the eradication or suppression of the released organism should the undertaking go awry. [8] Ideally, free-release guidelines should be international in scope. In these matters we once again encounter something of a parallel to the isolation of variolated subjects in colonial Boston. The spread of bioengineered organisms released into the environment must be monitored and in some cases perhaps limited, just as variolated subjects were monitored, and should have been isolated, by Boylston and Mather.

The use of genetically engineered life forms as weapons either by nations or terrorist groups is real but does not appear to be qualitatively different from the risks associated with traditional biological weapons. Nonetheless, continued study and investigation of this area seem warranted.

CHAPTER X

Bioburst
and Medicine

It may be surprising to realize that in the early days of this century testing of the chemical composition of blood to aid in the diagnosis of human illness was not only uncommon in medicine but in many quarters was actually frowned upon. Although today drawing blood for batteries of laboratory tests is an almost routine part of medical evaluation, some physicians not so long ago failed to see any value in measuring, for example, a patient's blood sodium or chloride level. After all, a good clinician can make the diagnosis of dehydration or acidosis from a careful evaluation of the patient and an examination of his urine. The blood tests in the eyes of these physicians added nothing except trouble and expense. Today it is widely appreciated that laboratory testing, including blood chemistry determination, is an indispensable part of medical practice. Those early opponents of blood testing did not realize that the chemical study of the blood could aid in the diagnosis of many diseases and provide invaluable assistance in the treatment of dehydration, acidosis, and a host of other conditions.

But the debate really was not about the value of determining the level of blood sodium or chloride in patients. The real issue was that some physicians schooled in the strictly clinical approach to medicine were intellectual xenophobes when it came to the terra incognita of laboratory medicine. They stood on one side of a philosophical gulf, modern medicine on the other.

Since those early days, the scientific advances produced by modern medical investigation have come rapidly, and present-day physicians accept continuing rapid growth in the technology and science of medicine as a routine part of the profession. But even many of these men may in the near future find themselves on the wrong side of an intellectual gulf as deep as that confronted by those physicians of yesteryear. The transition of medicine from a strictly organ-specific, disease-specific discipline to one that seeks out commonality in pathogenic mechanisms based on the precepts of cellular biology is inevitable. If the universe of medical knowledge began expanding with a big bang earlier in the century, now it will begin to contract—not because less knowledge will be available but rather because available knowledge will be sufficiently deep to permit the coalescence of previously disconnected clinical observations into coherent wholes. Stroke, heart attack, and vascular disease are seen in many instances to result from the same disease process. Cancer is becoming one of a small number of mechanistically similar disorders. Genetic diseases will yield to similar methods of diagnosis and gene therapy. Clinical specialists will, of course, be required in each branch of medicine, for diseases of the liver are not clinically the same as diseases of the heart. But the scientific basis for the treatment of diseases of each organ is likely to be the same.

The implication of this fact lies principally in the area of education. It is the standard policy of American medical schools to teach the scientific basis of medical practice. Instead, for example, of simply teaching that antibiotic therapy of streptococcal sore throat should be undertaken to prevent rheumatic fever, faculties routinely teach the autoimmune mechanisms by which certain infections can result in misguided immune attacks on the heart or kidneys. The better medical schools, however, teach not only the scientific basis of current medical practice, but also the science that will be the basis of practice a decade after the student graduates—and at the same time these schools inculcate in their students the unending need to stay abreast of progress in science. The students from these schools tend to find themselves among the first to arrive on the far side of any intellectual crevasse.

This method of pedagogy, characterized by its emphasis on the science of biomedicine, has certain definite applications to clinical care. This fact is often lost on medical students, who complain bitterly about learning scientific principles and facts when what they really want to do is immerse themselves in "hands-on" care of the patient. The oft-stated truth is, however, that the art of medicine is in many instances a matter of "making the right decision based on inadequate information." Medicine is a matter of judgment, and in a physician all the good intentions in the world cannot compensate for bad judgment. Awareness and critical evaluation of newly emerging scientific observations, however, probably can improve decision-making skills. Of necessity, the acceptance of this doctrine is something of an act of faith but one that can be rendered more palatable by example.

Twenty years ago, tremendous debate ranged in the medical community regarding the treatment of diabetes. Some research studies suggested that the vascular complications of this disease resulted from abnormalities that were inherited along with but not caused by abnormalities of sugar metabolism. Other research studies demonstrated biochemical changes in the nerves and kidney vessels of diabetic rats and further showed that these abnormalities improved when blood glucose was controlled with insulin. The implications of these contrasting data for the treatment of patients with diabetes are obvious. If the vascular complications of diabetes are independent of sugar, then it is pointless and perhaps risky to control blood glucose closely. If, on the other hand, vascular complications can be influenced by glucose levels, then efforts at maximizing glucose control are clearly warranted. Every physician caring for diabetics had to make a decision about the degree of blood glucose control he would seek in a given patient. Indeed, this debate still rages today, although less openly, and physicians are even now required to treat patients in the absence of perfect knowledge. Although many physicians now accept the notion that good control of blood glucose helps limit nerve and vascular complications, a recent article in the *Journal of the American Medical Association* reminds us that the issue is still debatable insofar as the prevention of diabetic eye disease is concerned.[1]

Questions of this sort can be resolved only by large clinical trials, and only such trials therefore can provide definitive therapeutic recommendations—but because of the enormous complexities of these undertakings even large nationwide trials often fail to achieve the goal of clarifying a difficult therapeutic conundrum. As an example, consider the apparently contradictory results of the Hypertension Detection and Followup Study (HDFP) and the Multiple Risk Factor Intervention Trial (MRFIT or "Mister Fit") on the effects of hypertensive therapy. HDFP concluded that the treatment of mild hypertension helped patients so afflicted, whereas MRFIT actually suggested that it harmed certain subclasses of patients. In the absence of conclusive trials, evaluation of scientific bench research is the only clue available to aid the physician in decision making. Some physicians twenty years ago may have given more weight to the set of experiments favoring tight control, while others may have favored the opposing data. Today some physicians may give credence to studies suggesting that certain antihypertensive drugs can actually increase the risk of heart disease in particular patient subgroups, while others favor the opposing studies. Nonetheless, it seems self-evident that the ability to review such data critically was, and is, an intrinsic part of making "the right decision based on inadequate information."[2]

If the emphasis on forefront research is pedagogically desirable in normal times, it is clearly much more desirable when a biological revolution is afoot. It would be prudent for all medical schools—including those that pride themselves on producing only clinical generalists—to emphasize the scientific basis of Bioburst. Failure to do this could produce pockets of medical illiteracy in the profession. The medical literature itself is beginning to change as an ever-increasing amount of molecular biology appears in clinical journals. The interpretation and proper use of this data requires a solid grounding in this new science.

The case for expanded education in these areas can be made even more strongly. The theory and practice of Bioburst should be taught extensively beginning in junior high school and continuing through college and graduate schools. This is evident when one realizes that business leaders, corporate executives, politicians, and investors will soon be confronted with the necessity of making informed decisions that relate to biological issues. Just as the entire educational establishment has risen to the challenge of the computer revolution with innovative teaching techniques and much enthusiasm, a similar approach should now be undertaken toward the teaching of molecular biology. We must become a population that is biologically literate.

Medical schools that are committed to the traditional curriculum must abandon their intellectual conservatism and wholeheartedly embrace the teaching of this new science. Although this may prove politically difficult at a time when the public is emphasizing primary care and the quality of social interaction between patient and physician, it should be pointed out that nothing in this proposal should be construed as detracting from the patient/physician relationship or the quality of primary care. Rather, education in molecular biology will help all physicians from the generalist to the superspecialist practice better medicine. It will enhance their appreciation of the mechanisms of health and disease, provide unifying principles, and help all physicians make the right decision based on inadequate information. And that is the role of science in medical education.

CHAPTER XI

Bioburst and the Pharmaceutical Industry

The implications of molecular biology for the pharmaceutical industry should be obvious. If stroke, heart attack, and vascular disease are conceptually and mechanistically related, then certain commercial realities are obvious. Consider, for example, a firm involved in the therapy of hypertension. Its research department may be heavily committed to research in traditional as well as novel areas of blood pressure control. In addition to adrenergic-blocking drugs, for example, it may be experimenting with serotonin antagonists and other innovative approaches to therapy. In ten years it may have a potent new compound available for marketing. But this effort and expense could be for naught if at the same time a competing foreign firm markets a compound that controls the development of vascular disease. Since it is through the production of vascular disease that hypertension does most of its damage, a drug that inhibits this process would likely limit considerably the need for potent new antihypertensive agents. In essence, the research activities of the pharmaceutical industry must transcend classical pharmacology and enter the world of cell biology.

Similar arguments could apply to drugs used in cancer chemotherapy as well as to routine laboratory tests for the diagnosis of infectious disease. If monoclonal antibody therapy should prove to cure leukemia, the market for traditional antineoplastic drugs will be reduced. If dot hybridization applied to micro blood samples will specifically detect viral and bacterial genes, the need for reagents used in traditional viral and bacteriological diagnostic cultures will be reduced. The current revolution in cancer therapy bears further witness to this point.

These essays have already touched on the idea that the immune system may play a role in the eradication of small cancers in man. And the central role of the T lymphocyte in the orchestration of the immune system has also been noted. Recently, scientists at the National Institutes of Health treated the lymphocytes of cancer patients with a growth factor called interleukin-2 in order to convert some of these cells into "lymphokine-activated-killer cells" (LAK cells) and then to induce these LAK cells to multiply. LAK cells have the capacity to attack a variety of cancers. Thus Dr. Stephen Rosenberg and his colleagues reinfused interleukin-stimulated LAK cells, along with massive doses of interleukin-2 to maintain their killing ability, into cancer patients and obtained promising results—the shrinkage of tumors—in some patients. Other research groups are actively investigating ways of producing LAK cells more efficiently, and even better results may be forthcoming. But in any case it should be noted that one of today's more promising approaches to cancer therapy is based on the cellular biology of the immune system and depends upon a growth factor, interleukin-2, which is only available in the quantities needed for the treatment of cancer because the gene coding for this protein has been cloned using the techniques of modern molecular biology. Other innovative approaches to cancer therapy are also firmly based in molecular biology. An intense effort is currently directed toward the identification of the genes responsible for cell differentiation—that is, the genes that cause a cell to assume a particular form and function in the body, or more simply, to behave normally. Scientists are attempting to determine if vitamins, drugs, or hormones can be used to turn on these genes in cancer cells so as to make them less cancerous, or even benign. Preliminary evidence seems to support the validity of this therapeutic concept.

These novel approaches to cancer therapy, like the innovative therapies soon to be developed against a host of other diseases, are firmly grounded in Bioburst technology. The pharmaceutical industry, like the medical profession, must prepare itself for these changes in market direction. Although these trends in the development of new therapeutic modalities clearly do not imply that all pharmaceutical firms must immediately embrace the Bioburst approach to drug development, they do nonetheless suggest that many of the promising therapies of the future will arise from the study of molecular and cellular biology.

For those pharmaceutical firms that have not as yet incorporated molecular biology into their operations, the first step must be the reeducation of research directors and corporate decision-makers so that the potential of molecular biology is fully appreciated. Each manufacturer will be compelled over the next decade to determine which elements of modern molecular biology affect his product area and, having done that, to establish research activities in those areas. These decisions necessarily will require that knowledge of the new science be widely distributed among management personnel. Considerable redirection of effort will likely be required over the next few years, and this will of course demand understanding and support at top levels of management.[1]

Firms can obtain the needed education in three ways. The first involves hiring people knowledgeable in the new technology to serve not only as researchers but also as in-house teachers. Larger pharmaceutical firms could find it convenient to acquire this expertise, along with production capacity, by simply buying a smaller biotechnology company. The second technique—the use of scientific consultants from academia—provides a wider base of perhaps more expert, but probably less committed (to industrial application), knowledge. The final approach is self-education by reading, attendance at seminars, and the like. These, of course, are the traditional modes of industrial education, but each is likely to be greatly expanded as the commercial applications of the new technology become increasingly appreciated.

Indeed, it may well be that providing useful information in this area will itself become a new industry, with newsletters, weekend courses, and the like springing up from universities and business schools around the country. Once the need for this kind of activity is fully perceived by the business community, these educational entrepreneurs will probably flourish.

Additionally, it is likely that a wide variety of consultative and collaborative ventures will be established between industry and academia, since it is unlikely that any one firm could afford all the in-house expertise it could profitably use in translating basic scientific discoveries into useful pharmaceuticals. Rather, pharmaceutical concerns will probably find it profitable to generate a core of in-house experts and researchers, who then judiciously collaborate with the basic science faculties of one or more universities. These in-house core groups likely would subserve two functionally distinct goals and for practical purposes could be considered to consist of two distinct components. The first of these, the working group, would consist of scientists engaged in the exploration of projects targeted by management as fertile areas for product development. These scientists would incorporate into their work the results of experimentation subcontracted to one or more university laboratories. Just as a mathematician often employs, in the construction of a new theorem, one or more lemmas or subtheorems, so too the working group would utilize one or more experimental "lemmas" developed in university laboratories to construct its final theorem or product. Project-related researchers in the working groups would thus serve as tangible points of contact between industry and academia.

Equally important, however, is the second component of the in-house core research group—the concept-development group. These scientists and research management personnel would be charged with following the biomedical and molecular biological literature to determine those areas of research most likely to yield novel pharmaceuticals. They, in addition, would be charged with identifying those academic groups whose collaboration would be required to achieve management goals. The concept-development group therefore would be responsible for developing successful strategies for product development based on the current state of biomedical and molecular biological knowledge. The working group, on the other hand, would be responsible for tactical implementation. Taken together, these two groups would provide industry with the ability to interact with academia in a near-optimal manner.

This approach will be good for industry and if properly established will be good for academia. It is likely that focused collaborative relationships between industry and academia could be constructed in ways that preserve academic freedom of inquiry and yet bring funds and ideas from industry to campus. Forging these arrangements will not be easy, but collaboration will come. Business-as-usual will soon prove inadequate in industry because of the changing nature of biomedical science, and in academia because of budgetary constraints placed on federal support of research. But it is entirely possible that the coming era of industrial-academic collaboration could prove to be a time of unprecedented productivity for both camps. The opportunities and challenges are clear.

There is one area of potential commercial-industrial application of Bioburst technology that cannot be inferred from our previous analysis of molecular biology and which, although clearly molecular, is at first glance only marginally biological. This is the area of digital nuclear magnetic resonance imaging (now often called magnetic resonance imaging), and as a fundamentally molecular diagnostic technique it appropriately should be considered part of Bioburst. To assess the implications, commercial or otherwise, of this technology requires that two questions be answered: "What is digital imaging?" and "What is nuclear magnetic resonance?"

Digital imaging refers to the visualization of body parts by computer techniques rather than by photographic film. The first major application of this technology was in the computer enhancement of TV pictures from distant space probes. In this case, signals from a spacecraft were not simply converted in a one-to-one fashion into dots on a picture; rather, computers using statistical techniques filtered out static and enhanced picture detail. The computer maximized the information in the spacecraft's signal—something that visual inspection of the pictures could never accomplish.

A theoretically similar though much more complex form of analysis gave rise to the CAT scanner, in which a patient is exposed to x-rays, but which provides a more complex analysis than simply recording the shadow of the beam on photographic film. The machine itself looks like a doughnut, the rim of which gradually moves around the patient. Rather than exposing the entire patient simultaneously to a wide beam, a fine x-ray beam is aimed from multiple directions at the body part being studied and the x-ray photons that pass through or are reflected from the patient are recorded by special counters arranged in a precise geometric relationship to the beam. From the tabulation of the innumerable x-ray photon counts scattered from the patient at each angle, a computer can construct a picture of great resolution. Body parts formerly difficult or impossible to study without the use of invasive techniques suddenly can be routinely visualized. Even strokes can be seen in brain tissues shortly after they occur. Moreover, this digital technology can be employed in conjunction with other imaging devices. Already, digital venous angiography is performing many of the diagnostic functions that previously were the exclusive purview of invasive arteriography. A simple venous injection of a contrast agent coupled with computer-image analysis can now in many cases provide as much information as a potentially (though rarely) dangerous and painful arterial injection. Indeed, the enhanced resolution provided by computer-constructed images coupled with the ease of data storage (x-ray films are bulky and flammable) make it inevitable that all medical imaging will soon be digitalized. The chest x-ray on a view box will soon be simply a vestige of medicine's quaint but primitive past.

In and of itself, however, digital imaging would not qualify as a Bioburst technology, but coupled with nuclear magnetic resonance (NMR), it clearly does. NMR is a technique originally applied in chemistry as a probe for determining the chemical structure of unknown compounds. It is based on the observation that (1) the nuclei of atoms consist of charged particles (protons) and uncharged particles (neutrons); and (2) these nuclei spin. It is a law of physics that spinning charges produce magnetic fields—or more precisely are associated with magnetic moments. Combinations of spinning charges may have a net magnetic moment or no magnetic moment at all, depending on whether the individual magnetic moments cancel out or not. Only the nuclei of certain elements—those possessing an odd number of protons or neutrons—possess net magnetic moments. NMR can *only* be used to study elements like hydrogen and phosphorus whose nuclei possess net magnetic moments—a minor drawback but one that must be kept in mind in projecting future applications of the technique.[2]

An NMR imager induces a strong homogeneous magnetic field through a large, three-dimensional slice of a patient's body, causing

135

many nuclei possessing magnetic moments in that slice to align with the field. This is a startling idea: billions upon billions of subatomic particles literally come to attention under the impressed field. The spin-axis of these nuclei rotate in such a field at a rate (frequency) that is dependent on their mass and magnetic moment, as well as on the strength of the impressed magnetic field. It can be noted that some of this rotation—more properly, *precession*—occurs even in the earth's magnetic field. Thus, each element possesses a characteristic precession frequency determined by its mass and magnetic moment.

These spinning magnetic nuclei are roughly analogous to a top spinning in a gravitational field. As every child knows, the axis of a spinning top rotates (precesses) slowly about a line perpendicular to the ground. So, too, magnetic nuclei precess about the direction of a magnetic field, and the rate of that precession is dependent on the mass and magnetic moment of the nuclei, as well as on the strength of the impressed field. If the impressed field is kept constant, the frequency of precession is then determined by the characteristics of the nuclei. Thus, each element possesses a characteristic precession frequency in a fixed magnetic field. If electromagnetic (*i.e.,* radio) energy of various frequencies is then applied to the nuclei, resonance considerations dictate that energy will be most efficiently absorbed by the nuclei at a frequency corresponding to their precession frequency. Thus, if one wishes to study the hydrogen in a sample, one would impress upon the sample—in addition to the fixed magnetic field—a radio frequency pulse at the characteristic frequency for hydrogen. This would disturb all the hydrogen protons to a degree dependent on the strength of the radio frequency energy applied, but would have little or no effect on other nuclei; the axis of spin of the hydrogen protons would, by virtue of the energy absorbed, be knocked out of alignment with the impressed magnetic field. If the radio frequency current is then turned off, the nuclei will return to alignment with the impressed magnetic field and in so doing will emit radio frequency energy (radio waves) as they give up the energy recently imparted to them (see figure 22).

The time required for the nuclei to return to the standard direction (as determined by NMR signal analysis) after a relatively brisk se-

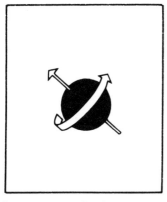

A proton spinning in the
normal magnetic field
of the earth

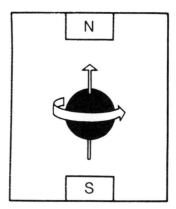

The same proton in an
NMR magnetic field

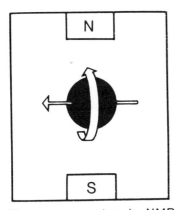

The same proton in NMR
after receiving radio-
frequency energy pulse

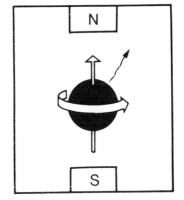

The proton returns to its
original position in NMR field
when radiofrequency energy
is turned off. In doing so,
it announces its presence by
emitting radio energy (⟶)

Fig. 22. Schematic representation of the participation of a proton in nuclear magnetic resonance imaging. Although the majority of protons align
parallel with the magnetic field, quantum mechanics permits a large minority to align antiparallel to the field. Strictly, it is the difference in the
number of parallel and antiparallel protons that generates a net proton
NMR signal.

ries of jolts is a characteristic of the element being studied. It can be measured by listening for the emitted radio signals and is called the *T1 relaxation time*. The strength of the signal emitted is related to the amount of the element present. If a somewhat different signal is analyzed, that obtained after a slightly different series of jolts, the second relaxation time, T2, is measured, and this is in part a function of other nuclei in the environs of the hydrogen nucleus being studied—that is, T2 is in part a function of the chemical compound in which the hydrogen nucleus finds itself. Thus it can be seen that the NMR radio signal, derived from listening to the radio waves emitted by these nuclei, contains a considerable amount of information. The frequency at which radio waves are absorbed and re-emitted by the sample provides information regarding the chemical element being studied. The T1 and T2 relaxation times contain information about the quantity of nuclei in the sample and their chemical surroundings. Because methods have been developed to alter intense magnetic fields over various parts of the body, and also to impress precise radio frequency signals rapidly over these specific areas, and because methods are available to measure the subsequently emitted radio signals, a computer can construct an image of the body part being studied.

In essence, the electronics of the NMR imager interrogate various parts of the body asking what their hydrogen density is. The images produced in this way are in most cases about as good as those generated by CAT scan, but for some body parts, such as the spinal column, they are substantially better. Because calcium does not possess a significant net magnetic moment, bone is essentially invisible to NMR. Thus, x-ray scatter from bone, the bane of all x-ray technologies, including CAT scanning, is eliminated. The spinal column can be studied thoroughly by NMR, but the calcium content of bone cannot be assessed at all. Additionally, lesions in the brain, including heretofore impossible-to-visualize demyelinating lesions like those associated with multiple sclerosis, are seen well. And finally, because cancerous tissue appears to be metabolically different from normal tissue by virtue of its water content and perhaps other factors, it can in some cases appear different from normal tissue in NMR imaging. Thus, the NMR can not only demonstrate a nodule in the liver but can suggest whether or not that nodule is malignant.[3]

There is still more to NMR. An obvious advantage to the technique, of course, is the avoidance of x-ray exposure. The patient is subjected to a strong magnetic field, but there are no known ill effects associated with this procedure, and there is theoretical reason to believe that it is entirely safe (the small amount of actual energy transferred to and from tissue during the imaging procedure is not much more than that associated with the radio frequency signal). A second major advantage of NMR imaging is the technique's potential to study other elements besides hydrogen. Because hydrogen is a part of just about all body components, proton (hydrogen) NMR is excellent for imaging. But phosphorus NMR could potentially be very useful for chemical studies of energy levels in tissues, since phosphorus is a major component of the body's energy source ATP. That is, NMR not only images, it has the potential to perform noninvasive chemical testing in the living patient. Already a patient has been diagnosed as having McArdle's syndrome (an inherited disease of energy production) on the basis of phosphorus NMR. Cancers, as noted, have sufficiently different chemistry from normal tissues to be in some cases diagnosed by NMR.

These observations also raise the possibility that therapy can be directed by NMR. Already efforts are under way to place chemotherapeutic agents in "microsomes" along with metal-containing compounds and to direct these microcapsules by focused magnetic fields to specific body locations. The extension of this work may permit the targeting of drugs to specific anatomic sites, such as widely spread tumor deposits. This approach potentially can permit the delivery of toxic agents to the tumors while sparing the rest of the body. Therapeutic agents in theory could be targeted to shore up weakened arteries (aneurysms) or to open plugged vessels. In short, this NMR-related technology could make possible computerized electronic microsurgery. The possible innovative uses of NMR technology therefore seem enormous. It has the potential for the selective imaging of body elements, the detailed study of metabolism, and even the possibility of directing microinterventions anywhere in the body.[4]

As opposed to traditional medical imaging modalities, such as x-ray or ultrasound, the conceptual basis of NMR technology is not scattering. An x-ray image is generated by the absorption or scattering (to positions off the direct line) of incident x-ray photons. CAT scanning (computerized tomography) carries this technique even further by assessing scattering directly. Ultrasound involves the scattering (reflection) of sound waves. NMR does none of this. NMR derives an enormous amount of information, including that needed to make a computer image, by first standardizing subatomic quantum conditions with the imposition of a magnetic field and then interrogating subatomic particles (nuclei) based on the quantum characteristics of their absorption of radio frequency energy. As such, NMR brings Bioburst to its ultimate extension—quantum chemistry.

In the final analysis, it is clear that the molecular biology of the gene, including as it does transcription, gene replication and crossover, mutation, and repair, represents a molecular—really quantum chemical—process. For the first time, biology is reaching the point where it can begin to appreciate molecular mechanisms. The nature of the interaction of a steroid-hormone-nuclear-receptor and DNA, the nature of the reaction between RNA polymerase, DNA, and other facilitating molecules, as well as other similar questions, are now within the reach of investigation. The next stage in understanding these processes will require quantum chemistry. Only this discipline will be able to tell us how likely DNA is to open in the presence of a specific hormone-receptor complex. Only this detailed kind of study will fully explain mutagenesis and carcinogenesis. The relationship of quantum chemistry to biological processes is the intellectual equivalent of the moon as seen from the earth. It is clearly there, but its potential has lain unrealized because of the enormous gulf (complexity) between it and biology. Nuclear magnetic resonance is a major step in the application of quantum principles to the real-time study of the living patient. It helps to close the gap. It is the lunar module of biology.

This brief description of NMR imaging is included in the discussion of Bioburst and the pharmaceutical industry for two reasons. First, industry has taken the lead in the development and financing of this technology, and second, industry will no doubt rapidly employ the fascinating properties of this modality in ingenious medical applications. However, this pattern of development may belie the theoretical implications of NMR technology, which in reality represents the palpable interface of quantum chemistry and medicine. Molecular biology has taken us to the cellular, subcellular, and macromolecular basis of medicine. NMR brings us to the natural termination of such an intellectual odyssey—to quantum chemistry. Any farther and physicians will again have to refer to themselves as practitioners of "physic."[5]

CHAPTER XII

Bioburst and
Employment Opportunities

As molecular biology comes into its own as a commercially valued source of drugs, diagnostic reagents, and agricultural products, it will create a large number of employment opportunities which, although not related to computer science or electronics, might still appropriately be described as "high tech." It is reasonable, therefore, to ask what academic preparation will be required to produce a work force of maximum productivity in this area. This issue is of obvious importance to industry, which must be assured of a supply of technically and intellectually qualified employees, and it is perhaps of even more importance to the student who hopes for a career in this emerging field. Society and industry must decide how best to cultivate aspiring practitioners of molecular biology, while students must decide how best to prepare for these careers. Superimposed on both these questions is the issue of how best to manage such extensively trained people once they are brought into a commercial enterprise. A traditional hierarchical structure would appear inadequate, at least in part because, except in firms starting *de novo*, newly recruited personnel are likely to be more knowledgeable in modern molecular biology than established senior management (including research management). Moreover, it is not at all clear that maximal productivity could arise from rigidly structured systems.

Perhaps the most important point related to education that can be deduced from this overview of modern molecular biology is that Bioburst technology is not an end in itself. Rather, it is a tool. Although one could make a career defining ways of maximizing reverse transcriptase production in a commercial setting, this activity only has meaning because of the potential applications of reverse transcriptase. Producing and bottling reverse transcriptase is not like producing and bottling soda pop. The consumer drinks soda pop—he can only use reverse transcriptase to make or discover something new. Thus, the power and creativity in modern molecular biology lie in its application and in the insights it makes possible to skilled investigators. Bioburst is a phenomenon that at once clarifies and expands other fields of science and philosophy and is utterly dependent upon these fields for its full fruition. This is not to say, of course, that profits will not be made from the efficient production of reverse transcriptase or protein hormones, as they will in numerous other support industries. This is already occurring. Nonetheless, the introduction of new diagnostic or therapeutic products is the area in which the pharmaceutical industry must compete, and in a wider sense it is the area, along with agriculture, that will determine the total productivity of the industry over time. If these innovative areas fail to live up to their promise, there will be little opportunity for growth in the support industries or for successful international competition in pharmaceuticals or agriculture. It is in this area of the creative use of molecular biology that broad-based, integrated knowledge is most critical.

The fact that Bioburst provides enormous insight but remains a tool of other disciplines has definite implications for training in this field. As much as the telescope revolutionized astronomy, its potential was only realized by men well grounded in the many branches of astronomy. The instrument was, and is, used very differently by planetary astronomers, solar astronomers, and deep-space astronomers. The lesson for training in molecular biology is that the student should first become truly literate in biology or medicine before undertaking detailed training in recombinant DNA technology if he wishes to advance in this field rather than simply to hold down a job.

This holds true even at the technician level. It is noteworthy that the technicians working today in rDNA-related fields are among the brightest young people in the country. Many are college graduates and have the ultimate goal of entering graduate or medical school. The value of well-trained technicians is made obvious when one considers that valuable applications and advances could potentially be lost or delayed if technicians failed to appreciate important observations made in the laboratory. Indeed, the productivity of industrial molecular biology, like the productivity of other newly emerging industries, will be critically dependent on well-educated personnel. As Peter Drucker, professor of social sciences at the Claremont Graduate School, wrote in the *Wall Street Journal,* "The switch to knowledge work as the economy's growth area and the large scale movement to the new technologies means above all that productivity will increasingly be determined by the knowledge and skill workers put into their tasks. And productivity, in the end, always determines the ability to pay and the level of real incomes. This is the reason we can say with confidence that the American school will improve—fast." [1]

The simple fact is that a continued supply of bright, well-educated "workers" is required by the Bioburst industries. Because we have no shortage of well-motivated young people (and that motivation will surely increase as the new economic realities are seen as irreversible), the real challenge is to construct an education system that will meet the needs of Bioburst science and related industries. Political considerations and professional squabbling must not be permitted to impede progress in this area. The new curricula must include solid grounding in biology, philosophy, mathematics, and the scientific method, as well as the detailed study of molecular biology. Such curricula most efficiently can be designed by representatives of the scientific, educational, and industrial communities. This kind of multidisciplinary planning of study programs may at first appear odd, but in fact we cannot afford the luxury of waiting for molecular biology to be "worked out" by scientists to the point that it can be homogenized and presented to students as home economics is. The cooperative construction of curricula for the junior high school, high school, and college years would seem to offer the best way of ensuring a supply of well-trained worker scientists in the future. This planning should begin now.

CHAPTER XIII

Aside:
Selfish DNA

When the techniques of nick translation, hybridization, and the like were applied to the mammalian genome, many surprises resulted. Jumping genes, introns, oncogenes, and similar phenomena sprang into the arena of science. And the concept of selfish DNA was born.

The deprecation of a molecule by the application of the term *selfish* is quite a startling idea. At first blush, it conjures up possibilities of an entire anthropomorphic chemistry. One can almost visualize "vivacious" glucose, "acrimonious" testosterone, "duplicitous" estrogen, and many others not fit for paper. The concept of "selfish" DNA raises the immediate question: if some DNA is selfish, is other DNA altruistic? At this point a tender chord is struck, because in a very real sense there may be altruistic DNA. All of us, in a way, may be made of altruistic DNA, since our bodies, our somatic cells, seem programed to die in the process of ensuring the survivability—indeed, the immortality—of the germ line. Still, altruism and selfishness are clearly anthropomorphic terms when applied to molecules, and it is an indication of the tremendous emotional and intellectual impact of modern molecular genetics that selfish DNA appears after a little thought to be a reasonable concept, while frivolous gonadotropin does not.

What is selfish DNA? Does it exist, and is it really selfish?

Recent studies of the mammalian genome, including that of man, have revealed the presence of hundreds of thousands of copies of various repeated sequences of DNA—some quite short, others thousands of base pairs long. These are liberally dispersed among functioning genes and for the most part do not appear to code for any specific functional protein. The revelation that a good percentage of human genomic DNA consists of the molecular equivalent of writing "I am a bad boy" on the blackboard three hundred thousand times, begged for explanation. Several were offered, including the concept of *selfish DNA*.

The suggestion was made that at least some of these repeated units—particularly the so-called intermediate repetitive sequences, which bear a resemblance to *transposons* (the jumping genes of bacteria)—are molecular parasites that benefited from evolution, be it molecular or macroscopic, without contributing to the good of any species. They code for no useful proteins and contribute nothing to the survival of the cell. The entire world of living things can then be viewed as unwittingly supporting these long repeats, which, having somehow got the genome to reproduce and protect them, simply sat back for the ride. Doing no useful work on their own, they were envisioned to be "selfish" DNA—the robber barons, if you will, of the evolutionary world or, better perhaps, the landed gentry of DNA, the old wealth of biology.[1]

After a little contemplation, this idea becomes rather unsettling. It flies in the face of common values and even appears to make a mockery of a good bit of Western morality. If life—human and otherwise—inadvertently works for the betterment of selfish DNA, then so do humanistic values such as altruism, hard work, and honesty. One can almost see the repetitive DNA slowly expanding and contracting—the DNA equivalent of chuckling—over this development. Altruistic human beings working for the betterment of a lazy bit of DNA might be amusing to the DNA but could be troubling to the moralists. For if it can be argued that Newtonian mechanics provided the basis for a deterministic social philosophy, that Darwinian evolution provided a climate hospitable to the flourishing of economic "natural selection" in the form of capitalism, that quantum mechanics reintroduced doubt, uncertainty, and unknowability not only into physics but into the affairs of man, then might not the demonstration of quintessentially selfish DNA generate a social tidal wave? Has the epoch of the hedonist truly arrived?

However, the emotional and social threat posed by selfish DNA to humanistic values may yet be removed. The repeated sequences may in fact have important functions after all. There is evidence to suggest that some of them may influence the expression of nearby genes while imparting a sense of structural organization on the entire genome. This structural or organizational role may be particularly important during the complex process of cellular division. No one knows for certain, but this seems a possibility. And these sequences may subserve other roles still to be defined. Indeed, some of this DNA may be the stuff from which a good deal of evolutionary progress is fashioned. It is possible that selfish DNA is not selfish at all.[2]

This state of affairs potentially removes the difficulty of trying to explain what evolutionary pressure could be invoked to permit the maintenance of useless DNA in the genome, and it confirms our intrinsic belief that nature, and more particularly evolution, is efficient and not about to be taken in by a molecule whose crime is not so much selfishness (which is a trait that can appear in many life forms) but useless selfishness—in that the selfish entity is of no use to the host genome or to any form of life other than itself. The idea that selfish DNA may in some way be functional is perhaps more pleasing to our preconceptions about values and nature.

Yet the idea that repeat sequences serve some function could be incorrect. Stephen J. Gould, for example, has argued that on the molecular level carrying a *limited* amount of useless excess baggage could be evolutionarily silent, having no adverse impact on survival. There is no theoretical reason why selfish DNA cannot exist.[3]

Is some DNA "selfish"? "Uselessly selfish"? To our anthropomorphic minds, a morality play is being enacted in the molecular world. We can only wait for the final curtain to come down. But even if some DNA is shown to be uselessly selfish, its existence may in the end teach us less about the work ethic and more about charity. For selfish DNA, useless as it may be, is nonetheless part of us and of life, and as such it is hard to be severe in judging it. More important, the existence of selfish DNA would likely teach us once and for all that human morality derives from the heart, mind, and history of the animal that is man and not from universal principles of biology.

CHAPTER XIV

Bioburst and Aging

It has been said that the only constants in life are death and taxes. Although molecular biology cannot reasonably be expected to eliminate taxes, it is not unrealistic to ask if it can prevent or delay death. This, however, becomes a complex question because there are at least three major mechanisms by which people die. The first of course is disease—in which case an easily definable entity, often a microbe, produces the failure of one or more specific body parts. The second is accident, and the third is aging. It could be argued that aging doesn't truly kill but simply increases the likelihood of fatal disease or accident; but since the octogenarian who contracts simple pneumococcal pneumonia (or other usually curable disease) often dies while the teen-ager lives, it is reasonable to consider aging per se as a mechanism of death.[1]

Molecular biology has an enormous potential for the control of disease, so it will undoubtedly prolong lives. It can do little about reducing accidents and so cannot provide true immortality—that is, anyone who lives long enough will eventually be hit by a truck or experience a comparable mishap. Stochastic reality is not consistent with physical immortality.

But the big question is: does molecular biology offer the possibility of retarding aging so that, accidents notwithstanding, the human life span could increase to say, three hundred years? In order to answer this question, it is necessary first to decide what exactly aging is. Let us somewhat arbitrarily define aging as that process, intrinsic to the normal functioning of the organism, which results, after an organism reaches maturity, in a progressive decrease in biological functions, including those functions related to viability or, as we shall call it, "longevity." Let us, again quite arbitrarily, define "longevity" as the probability that the organism will be alive at some predetermined time in the future—say, for simplicity in the human case, one year. This definition of aging is far from perfect, since social context can in many ways offset or exacerbate the effects of aging on "longevity." For example, an animal who with age loses his ability to run and therefore hunt in the wild might soon starve to death, while a similarly afflicted animal fed in a zoo might live considerably longer. Yet it cannot be said that the animals "aged" at different rates. Perhaps what we really seek is the probability that the animal would be alive under "ideal" environmental conditions. The definition thus becomes somewhat nebulous, but it is sufficient to make clear the essence of the process we wish to describe as well as the ambiguities inherent in the definition itself. One additional point should be made. While accidents decrease longevity, they do not do so gradually. Aging decreases longevity gradually. (Additionally, it is possible that different tissues age at somewhat different rates in different individuals. There may in fact not be any one index of "biological" age. It may be more appropriate to consider organ- or tissue-specific aging.) A plot of longevity versus time for a mythical nonaging person would look as shown in figure 23. A similar plot for an aging person might appear as shown in figure 24.

Aging, then, can be viewed as a gradual decline in longevity coupled with a decline in biological function, the loss of biological function being assessed in any number of ways, such as the ability to run, to think quickly, and the like. While the average life expectancy of men in the United States has risen from about 47 in 1900 to 73 in 1977, the maximum human life span has probably remained constant at about 120 years for thousands of years. In other words, although many more people live longer today than they did in the past, the *oldest* living persons in Roman times were probably the same age as the oldest living people today. This suggests the existence of a more or less immutable biological clock that controls senescence in man.[2]

If this deduction about the existence of a biological clock is cor-

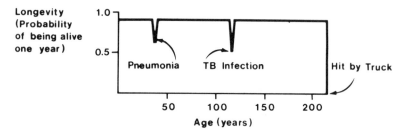

Fig. 23. Plot of longevity versus time for nonaging person.

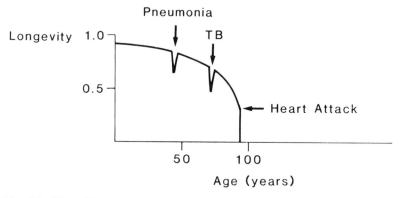

Fig. 24. Plot of longevity versus time for aging person.

rect, the next logical question is: where does it reside? Of course, there is no a priori reason why the fundamental mechanisms responsible for age-related organ failure must necessarily be the same in all tissues. There are those who believe that one or another organ in the body, such as the pituitary gland, for example, possesses the biological clock and, by virtue of hormonal signals sent throughout the body, orchestrates senescence. Some experimental data actually support this hypothesis. If this theory should prove to be even partially correct, it would auger well for therapeutic retardation of aging, since it would appear to be at least conceptually simple to isolate the responsible hormonal signals and then to "correct" their levels in blood to a more "youthful" value.[3]

Bioburst and Aging

Unfortunately, considerable experimental data can be arrayed against the idea that an organ-based biological clock is the *sole* basis of aging. Although active research on the organ model is under way, the focus of scientific inquiry is turning to the cellular level. This interest in cellular aging began with the controversial studies by Hayflick and his coworkers. These scientists demonstrated that mammalian cells, including those of man, can be passed in tissue culture (that is, transferred from dish to dish as their growth fills each dish) only a limited number of times before they simply stop growing. To reiterate, if one allows cells to grow in culture dishes until they fill the dish and then removes them, taking a fixed number to populate a second dish, and then repeats this process, after a fixed number of passages from one dish to another (usually about fifty the way the experiment was originally set up) the number of cells capable of growing in the new dish gradually decreases to zero. Cells taken from older people or animals appear to reach their so-called mitotic limit after fewer passages than those from younger people or younger animals, although this finding is contested. Thus, these data seem to imply that mammalian cells are capable of only a fixed number of cell divisions in their lifetime, and that the cell itself contains the biological clock that orchestrates its senescence.[4]

These experiments were challenged on several grounds, which might be summed up by saying that cells in tissue culture passed from dish to dish are not comparable to cells in the intact animal. Tissue culture conditions could select for less viable cells, for example. But these arguments notwithstanding, the observation is increasingly appreciated by the scientific community as a valid one with important implications for aging.

If, as appears to be the case, cellular aging lies at the heart of human senescence, is there anything that can be done about it? In order to begin to answer this question, one must recall that not all living cells senesce. For example, single-cell organisms like the bacterium *E. coli* divide regularly into daughter cells and continue to do so until accidental death intervenes. The *E. coli* living today are, as best anyone can tell, the direct descendants of the *E. coli* of millions of years ago. *E. coli* in this sense are immortal. There is no obvious senescence of *E. coli* either. The *E. coli* of today, we have every reason to believe, are as vital and functional as the *E. coli* of a hundred or a thousand years ago. Were this not the case, if *E. coli* truly experienced senescence, *E. coli* would become weaker with time and would be expected at some point in the future to die. There is no evidence of this senescence after over seventy years of study of these organisms. We may resonably conclude that they are indeed immortal.

However, although the *E. coli* of today are the direct descendants of the *E. coli* of the past, they are not identical to them. In the intervening years, mutation of DNA and sexual conjugation, with a resultant admixture of DNA to the line of any one *E. coli,* have occurred. It remains an interesting question whether an *E. coli* strain devoid of the ability for conjugation and protected from mutation or DNA rearrangements could remain immortal. Some simple organisms have been reported to exhibit immortality without sexual conjugation; thus, sexual genetic mixing does not appear to be a stringent requirement for immortality, although this question should probably be reinvestigated using modern techniques of molecular biology.

In the case of humans and higher animals, a similar immortality can be found, but in this case the immortality resides in the germ line. It is self-evident that all humans are the issue of progenitor humans and that the lifeline of each of us could in theory be carried back in an unbroken chain to the dawn of life. The germ cells of humans essentially live forever, as they must if the species is not to die out. Additionally, there is no evidence of senescence in the human germ line. The sperm and ova of today appear to be as vital as the sperm and ova of the past. Confirmation of this assertion is found in the fact that the product of today's sperm and ova—today's children—are as vital as the children of any past age. The human germ line, like the *E. coli* or cancer cells in tissue culture, is immortal. It is only the somatic phenotype—the human animal, which the germ cells use in their effort to propagate—that senesces. Parenthetically, it can be noted that the infusion of new genetic material is an intrinsic part of germ line life.

One could argue that the senescence and death of the individual human is irrelevant to, or even possibly advantageous to, the germ cells. Traits that enhance survival during the early reproductive years might well have been "selected for" by evolution even if they caused deterioration of the individual in later years. It is also possible that the death of "early" organisms reduced the competition faced by the new products of germ line mixing in the battle for survival. These are potentially important, albeit speculative, points, for they suggest mechanisms by which senescence might have arisen and also lend credence to the idea that attempts to tamper with senescence may have serious consequences for the species.

It is well established, for example, that the birthrate in the United States has recently declined, while the number of elderly, thanks to modern medicine, has increased. In 1983, for the first time in the country's history, the number of elderly exceeded the number of teen-agers. There are many reasons for this at least temporary decline in the country's birthrate, but one likely reason is the poor economic climate of recent years. Indeed, it could be argued that one factor tending to produce an economic climate less hospitable to childbearing is the transfer of resources from the young to a growing number of elderly. If there were no elderly, more resources would be available for the support of children. All other things being equal, there would likely be more children. This is not to say that society should not support the elderly or that transfer payments to the elderly currently are a major cause of the economic difficulties of young families. There are other transfer payments—defense spending, and other social spending—that have a similar effect on the income of small families. Nonetheless, the transfer of wealth from the young to the old is a real phenomenon and, as the number of elderly increases, will probably become economically more important.[5]

If it is true that an increase in life expectancy to 75 years can have an impact, however small, on birthrates, how much more impact would an increase in longevity to 140 years have? Any attempt to answer questions of this sort must depend on estimating such variables as the health of the aging population, the availability of critical resources, and the general vigor of economic activity. Putting aside all economic arguments with their many confounding variables, what can one reasonably expect the birthrate to be if people essentially lived forever? Considerations of food supply and space would quickly eliminate the possibility of rearing children. Just as Draconian measures are currently being applied in the People's Republic of China to limit the birthrate in the presence of excessive population, so, too, would all societies, in the absence of aging, be compelled to utilize such measures. That could spell the end of the species as we know it. *Homo sapiens* could lose the vitality and resiliency that comes from constant genetic mixing and renewal. Selective pressures would all but disappear and the species would be converted to a breathing fossil frozen in time. The absence of senescence, coupled with vigorous efforts to eradicate disease, would dramatically reduce natural biological variability over time unless new worlds were available to absorb the nonaging, long-living humans and thereby make room for new generations of new genetic composition. Fortunately, NASA is about as close to major colonization of other planets as molecular biology is to slowing aging. One could also ask if the effects of continued biological selective pressures are actually essential to the ultimate well-being of the species. If the human species could be forever frozen in time in its current state, the genetic engineer would then serve the role of evolution by continually altering the genomic material of living humans to permit them to fend off newly evolved organisms, viruses, and other pathogens. Although (perhaps for reasons bred into the human brain by the biology of senescence) this prospect does not seem very appealing, it nonetheless is one that should be considered.[6]

Several important points can be made regarding the observation that not all cells age and yet somatic human cells do. First, we can conclude that senescence is not a strict biological necessity, since not all cells age. Second, as pointed out by August Weismann in the nineteenth century, aging appears to occur in the biological world only among those organisms that have developed specialized organs for reproduction and others for the day-to-day business of living. That is, it seems that aging is a characteristic of life forms that consist of an (immortal) germ line and a (mortal) soma. How this came to be is unknown.

One can next ask the more general question: if some cell lines are immortal and others are not, is this the result of the absence of a "juvenile factor" or the presence of "aging factors" in the mortal cells? This question cannot yet be answered definitively. It has been shown that with the passage of time pigments accumulate in vacuoles within aging cells. These appear to represent the accumulation of cellular components irreparably damaged during the processes of cellular metabolism. Could these changes lie at the heart of the aging process? Or could it be that random damage to DNA produced during normal cellular functioning is not perfectly repaired, leading to an accumulation of gene defects? Both these hypotheses provide possible avenues of exploration and both lend themselves to experimental testing. The important point is that the elucidation of the mechanism of aging may be possible in the not-too-distant future, and with this knowledge will likely come insights into the possible retardation of the process. Indeed, the retardation of aging, based on what is now known, in theory seems achievable. There is no theoretical reason why one or more interventions could not be discovered to protect the genome or other important cellular constituents and thus retard aging. In fact, recent studies suggest that cellular immortality, whatever its mechanism, is a recessively inherited trait. This in turn suggests that a genetically directed process produces aging. If cells produce their own aging, then perhaps the simple interruption of this active process would prevent aging. This would theoretically appear to be an eminently achievable result and perhaps the most optimistic scenario for the retardation or elimination of aging discussed thus far. Just as one can envision genetic therapy capable of correcting the abnormal genes responsible for thalassemia, so, too, it is possible to envision therapy directed against the genes responsible for senescence. On the other hand, the elimination or significant retardation of the aging process may require detailed interaction with genomic DNA with a precision that is forbidden by the laws of physics. And it remains possible that senescence derives from the impairment of some other cellular function(s) or that it is indissolubly linked in some fashion to normal cellular differentiation. Thus it will be impossible to determine the feasibility of eliminating senescence from the human species until more is known about its precise molecular mechanism at the most fundamental level.[7]

But it might still be wise for NASA to speed up its space exploration program.

CHAPTER XV

Information, Statistical Mechanics, and Human Aging

The detailed description of DNA is in essence a problem in biological chemistry. If, as seems reasonable, aging derives from alterations in DNA, then biological chemistry—and, in the final analysis, quantum chemistry—must play a role in understanding this process. But no complete description of the biological chemistry, much less the quantum chemistry, of aging is available, and if it were, it would likely be extremely complex. On the other hand, one can appreciate the nature of senescence by taking a somewhat more oblique tack.

That considerations based on statistical or quantum mechanics could limit intervention into the process of aging perhaps can be most easily appreciated when one considers that information represents free (that is, usable) energy. More particularly, the information contained in DNA, cell architecture, and tissue organization represents a considerable amount of usable energy. At the same time, information represents the improbability of messages or patterns occurring on the basis of chance alone. Thus the informational content of a cell is inversely related to the probability that the cell could spontaneously form from its constituent parts—while also providing an index of the amount of energy required to describe the structure of the cell.

These conclusions can be inferred from communication theory—or, more generally, information theory—which was developed in the 1940s by Claude Shannon. The origins of this discipline are found in a classical thought experiment described by the renowned physicist James Clerk Maxwell. Maxwell realized that, as required by the second law of thermodynamics, heat never spontaneously flows from a cold to a warm object, and yet he knew that the temperature of a gas is determined by the rate of motion of the molecules composing the gas. The faster the gas molecules move through space, the hotter the gas. The random motion of the gas molecules of course results in a distribution of speeds of the individual molecules, but the average speed is greater in a hot gas than in a cold one. Maxwell argued that if a cold gas were placed in one chamber and a hot gas in an adjoining chamber, heat energy could be made to flow from the cold to the hot gas if a microscopic "demon" sat at a door connecting the two chambers and opened the door when a fast-moving molecule in the cold chamber approached the door. If the demon similarly opened the door when a slow molecule approached from the hot side, the net effect would be to increase the average speed of the gas molecules in the hot chamber and to decrease the speed of those in the cold chamber. That is, heat energy would have flowed from the cold gas to the hot—in violation of the second law of thermodynamics.

For several years, a reconciliation of this thought experiment with the laws of thermodynamics eluded scientists, thereby giving the "demon" his name. Then Leo Szilard pointed out that the demon could not perform his magic by randomly opening and closing the door. Rather, he required information about the speed of approaching molecules. According to Szilard, this information was the equivalent of free energy. If the information-energy required by the demon were equal to the minimum amount of energy required by, say, a refrigerator that further cooled the cold gas and warmed the hot gas with its exhaust, then the thought experiment would be reconciled with thermodynamics and statistical mechanics. The second law of thermodynamics after all doesn't say heat *can't* be transferred from a cold object to a warm one, only that considerable energy input (as in, for example, a refrigerator) is required to perform this feat.[1]

Considerations such as these led to Claude Shannon's development of communication theory. That this theory is related to thermodynamics is obvious from the story of Maxwell's demon. Since this thought experiment involves the description of molecular events, it is likewise apparent that communication theory touches on quantum mechanics.[2] This latter relationship is reinforced by the fact that Shannon's work lies at the heart of the controversial "many-worlds interpretation" of quantum mechanics—a topic that will be considered later. Communication theory is a simple yet enlightening tool that potentially can serve as a framework for analyzing, in broad terms, the process of aging even in the absence of a detailed chemical model of senescence.

Shannon defined the informational "entropy" of any source or transmitter used to convey information as a measure of the uncertainty that must be compensated for by information if accurate communication is to be achieved. More specifically, he demonstrated that informational entropy could be quantitatively calculated from the apparently simple formula:

$$H = -\sum p_i \log_2 p_i$$

where H is entropy, p_i is the probability that any given symbol in one's alphabet could occur in the message position under consideration, and the sum is performed over all letters in the sender's alphabet.

To examine this idea further, consider the following sentence: "Jack and Jill ran up the hil-." One can ask, What is the informational content of the blank last symbol of the message? If any letter could with equal likelihood fill this slot, the probability assigned to each letter that could fill this last slot would be one divided by the number of letters in the English alphabet, or $\frac{1}{26} = 0.04$. Then

$$H = -\sum_{i=1}^{26} (.04)_i \log_2(.04)_i = 4.83 \; bits \text{ of information.}$$

This is an estimate of the amount of free energy needed to specify the last letter of the sentence.

However, the actual information content of this symbol is substantially less than 4.83 bits because in English the letters $A,B,C,D,E,F,$ $G,H,I,J,K,M,N,O,P,Q,R,S,U,V,W,X,Y,$ and Z cannot be used to fill this slot. Really, only L and T can be inserted in this blank to make a common English word. Therefore, $p = \frac{1}{2} = 0.5$ and

$$H = - \sum_{i=1}^{2} (.5)_i \log_2 (.5)_i = 1 \text{ bit.}$$

This is considerably less than the information requirement calculated previously, and this reduction in informational content derives from the fact that, in context, the last symbol in our text sentence is highly constrained. Indeed, the informational content associated with this last symbol is even less than we have calculated (could Jack and Jill really have "run up the hilt"?). "Jack and Jill ran up the hill" is almost certainly the message ($P=1$, $H=0$). But in that case does the last symbol really tell us anything? Is it worthwhile in transmitting this message to a distant outpost by radio even to expend the energy needed to transmit the last symbol? Could not the message "Jack and Jill ran up the hil" have done just as well in conveying the information that we wish to transmit?

This example further illustrates the relationship between energy and information. At the very least, it demonstrates that it is possible to waste energy in the transmission of information. At the same time it suggests that there is a certain minimum amount of energy required for the transmission of any message against the all-pervasive background of thermal noise. Shannon calculated that the absolute power required as a minimum for the transmission of C bits of information/second at temperature T above absolute zero is given by:

$$P \text{ (in watts/sec)} = 0.693 \text{ } kTC, \text{ where}$$
$$k = \text{Boltzmann's constant or}$$
$$1.37 \times 10^{-23} \text{ joules/degree celsius.}$$

In contemplating this formula, one might be struck by the relatively low free energy content of information. After all, our formula tells us that at room temperature, the transmission of one bit per second of information requires less than 10^{-20} watts. But it is important to recall that the energy we have calculated represents the theoretical *minimum* energy required to transmit the information. The amount of information actually required to describe biological processes is extremely large.

In order to elaborate on this theme, consider now the same sentence, "Jack and Jill ran up the hill," but assume some atmospheric disturbance during the radio transmission of the message—what is technically called a "noisy channel." When transmitting information over a noisy channel, one may be required to transmit the same message many times before the sender can conclude with reasonable certainty that the correct message has been received. For example, if the first transmission ran "Jack [*static, static*] ran up [*static, static*]" the meaning of the sender would not be clear to the receiver. This message could easily read, "Jack (but not I) ran up (the bill)." Context is not adequate to make meaning clear. With the next transmission, the following might be received: "Jack [*static*] Jill [*static*] up the hill." Taken together with the first transmission, the message is now clear. But because of the noisy channel, the message had to be sent twice—that is, twice as much energy was required for transmission. Thus it can be observed that for any message there is a minimum amount of energy required to transmit it on any channel, and the necessary energy is determined by the informational content of the text and the noise of the channel.

What has this to do with aging? Potentially, a great deal, for it permits us to discuss scenarios of aging and strategies for intervening in the aging process without knowing the details of the process or actually getting into biological or quantum chemistry.

To see this potential applicability, consider the preparation of a dish of scrambled eggs. First, an egg is broken against the rim of a frying pan, shattering its shell into hundreds of random pieces. If one were to stop at this point and attempt to reconstruct the egg, the job would be considerable, for although there is only one way in which the shell fragments could be reconstructed to produce the eggshell, an enormous quantity of information would be required to categorize each fragment as to size and shape, and then to determine how they all fit together. And because there is only one way to put the eggshell back together exactly as it was, there is no margin for error in the transmission and storage of that information. Thus the energy requirement for the accomplishment of this task is substantial. Consider now the next step—scrambling the egg. If midway into this process one wished to reconstruct the egg, how much information would be required to replace globs of yoke in just the right place? And what about unmixing mixed molecules, not to mention molecules that are constantly subjected to thermal motion? Because each step of the unmixing process likely would require multiple attempts before arriving at a successful conclusion—by virtue of the probabilistic nature of the microscopic world—much information would be required before the success of each phase of the unmixing could be confirmed. The information, and hence the information-associated energy, required for the unmixing and subsequent reassembly is so enormous as to convince us that the operation is, for all practical purposes, impossible. And the large informational content of an egg implies that the egg is not likely to unscramble spontaneously. No one has ever unscrambled an egg, and no one ever will. This is yet another way of approaching the chemistry of the reaction—we are using information theory to arrive at approximate answers to questions involving the reconstitution of structures. Of course, this is not to imply that information theory "explains" or replaces the second law of thermodynamics. In fact, the second law was used in the initial development of the equations of information theory, and information theory is used here only to make more palpably real some of the implications of the second law.[3]

An objection could be raised at this point. If the information required to make an egg is so great, how do chickens perform the trick? Or, to be more explicit, if a chicken were to be fed scrambled eggs, would not some of the chemicals in the scrambled eggs wind up in the eggs newly laid by the chicken? Has not the chicken, in part, done the impossible?

Not really. First, there is a great deal of information required to make a chicken egg, and it is something of a wonder that the energy provided by the metabolism of foodstuffs in the form of ATP is sufficient to accomplish this feat. But remember, the egg is made from a genetic blueprint (which developed, utilizing energy input over millenia, by a process that remains a mystery) contained in its DNA as well as in all the information found in the germ cell structure of the chicken. A great deal of the information needed to make an egg is already available to the chicken. This information-energy is not available to the chef who tries to reconstitute an egg. Of course, the energy required to make an egg is attainable since chickens can lay eggs. But in the absence of a genetic blueprint and an extraordinarily complex machine for making an egg in the absence of the chicken, the information needed to direct this kind of construction is mind-boggling. Even more important, one must notice that the chicken makes a new egg, not the egg that had been scrambled. The difference between *a new egg* and *the egg anew* is the difference between the possible and the impossible, because not even the chicken can lay a second egg identical to the first.

What relevance this kind of analysis has to the biology of man can be made clear when one considers the patient who has recently suffered a stroke. The brain can be perceived as a collection of millions of nerve endings, each abutting upon a specific part of another brain neuron. Although there is "plasticity" in these connections at various stages of development, the brain at any given moment is in one particular state or another. If a stroke destroys some of that wiring, the information required to rewire the circuits exactly as they were before is so enormous as to make the task impossible. The functional plasticity of the brain can sometimes allow signals to be rerouted so as to restore some lost function, but a complete return to the prestroke state is not possible. (This is one reason why the prevention of vascular disease leading to stroke is so important.)[4]

This example of stroke demonstrates that information theory may have implications for medicine. However, it is important not to overinterpret these lessons. Brain cells do not normally multiply in the adult, so cells lost to a stroke are lost forever. Perhaps the provision of a growth factor to the area of the brain affected by stroke would lead to neuronal reproduction and the replacement of damaged cells. Additionally, the injection of fetal brain cells might potentially replace the lost cells and restore some neurological function, as experiments in animals with various neurological disorders have demonstrated. Although the original connections cannot be exactly restored (by analogy with the impossibility of unscrambling an egg), an acceptable functional result (movement of limbs, ability to speak, and so forth) could result from growth factor or fetal brain cell administration.

But what about thought patterns, memories, personality? If, as appears likely, these things are based upon diffuse neuronal "wir-

163

ing" networks, then some changes must occur following any widespread cortical rewiring. Whether or not these changes would be perceptible to the stroke victim or his associates is an open question. It may be that, although exact rewiring has not occurred, the patient functions and feels perfectly normal. Although information theory quite properly indicates that we cannot rewire the brain perfectly, from a practical point of view this could be of little consequence to the patient's functional status.[5]

Thus, although information theory can tell us what we *cannot* do, it does not tell us what we need do to achieve a result acceptable to us. For example, the issue of brain function and information theory is potentially more complex (and in ways that begin to impinge on our views of aging) than we have already described. What if the *engrams* (the hypothesized storage form of memories in the brain) wear down with time? What if the systems for maintaining engrams fail over a two-hundred-year period, just as senescence mechanisms result in body failure after a maximum of about a hundred and twenty years? In this case, an intervention that stops senescence indefinitely would result in people who lost their personalities—their essence, their souls—after a couple of hundred years. Even if they could begin to relearn (form new engrams) again after that time, would they be the same people? Or would they have "died" in spite of their somatic immortality? This remains only a hypothetical question, for no one knows what the maximal lifetime of the engram is, since experience can only tell us that the engram lasts as long or longer than the human life-span. An upper limit on the length of survival of the engram cannot be deduced or surmised from any observation made thus far.[6]

In sum, it can be seen that general conclusions regarding the likelihood or even the possibility of certain outcomes can be derived from information theory, but detailed predictions in any specific case can only be based on detailed quantitative information and quantum mechanical analysis regarding the system under study. As a general rule, this kind of information is unavailable in the case of living systems.

Applying this kind of questioning directly to the aging problem, we can then wonder, If multiple minute molecular changes in DNA (or in any other irreplaceable yet critical cellular component), perhaps resulting as a natural part of the transcription or reproductive activities of genes and/or from free radicals generated by metabolism produce changes over long stretches of DNA (or other cellular components), might it not require so much specific information to correct them as to be infinitely more favorable energistically (and probabilistically) to build a new human being than to keep an old one unscrambled? Could it not be that the vast amount of information needed to locate and precisely correct the damaged molecular structures makes these tasks impossible? (So, too, might not a similar argument demonstrate that it is theoretically impossible to preserve or repair the engram?) If so, death and taxes necessarily will last forever—by fiat of thermodynamics and statistical mechanics. The correction of damaged DNA would then properly be considered roughly equivalent to unscrambling an egg. It would be impossible.

On the other hand, the immortality of lower life forms such as bacteria, like that of cancerous or transformed cells in tissue culture, suggests that the problem may not be so formidable as the above hypothesis suggests. The information (energy) demands may not be as great as feared. For example, if a relative deficiency of a DNA excision-and-repair system accounts for the rapid aging of certain species by permitting the accumulation of errors in DNA sequence or structure, then the problem is considerably simplified. In this case, simply raising enzyme levels in early life (and perhaps even in later years) could have the beneficial effect of stopping or slowing aging. The aging problem becomes soluble because the informational burden associated with preventing aging is not inordinate and has, in large part, already been borne by the natural development of the enzymes comprising the excision-repair system. The information needed would be only that contained in the DNA coding for the enzymes and for those techniques needed to insert that DNA properly into the target cells. Science would not be starting from scratch.

Strictly speaking, the total information content of the human genome, calculated as

$$H = -10^{10} \sum_{i=1}^{4} P_i \log_2 P_i$$

where $P_i = \frac{1}{4}$, is small in energy terms when one considers the previously mentioned formula $P(\text{watts}) \cong 1.37 \times 10^{-23} \, CT$. This calculation of the information content of the genome assumes (a) that the nucleotides are available as preformed building blocks so that there are four choices for each position in DNA; and (b) that there are about 10^{10} nucleotides in the human genome. However, this is a minimum estimate. The information required to direct either the synthesis or the extensive repair of the genome could be enormously greater depending on what aids are available (*i.e.*, what cloned genes for exogenous polymerases or what templates are available). To do this chemically from scratch in a living cell without any enzyme helpers would almost certainly be informationally impossible. However, this is not the problem with which we are confronted, since enzymes and the like are potentially available to us.

Even if the information needed to eliminate senescence should prove unachievable, that needed simply to retard aging may be (indeed, based on experimental studies in animals, almost certainly is) achievable. An intervention as simple as the avoidance of foods known to produce high concentrations of free radicals upon their metabolism could theoretically be a mode of retarding aging if the DNA damage hypothesis as a cause of senescence is correct. Even more effective and direct approaches to the retardation of senescence remain energistically possible. The retardation of aging would be a major accomplishment, and even the elimination of senescence may be possible. We are very close to knowing for sure.

CHAPTER XVI

Bioburst and Evolution

There is a storm enveloping classical biology that could have profound implications for social theory and policy. It has been building for over one hundred thirty years and involves a major pillar of scientific thought and process. Darwin's theory of evolution is under attack from many quarters, and the infighting has become intense. Indeed, this quiet battle with potentially important consequences for social philosophy has spilled over even into the weekly news magazines. Although this debate does not directly involve Bioburst, it is possible that Bioburst can shed some light on the discussion that is taking place. In addition, it is important to the appreciation of all life sciences, including molecular biology, that those terms and concepts that have come under question be clearly defined and distinguished from those that remain sound.

It will be recalled that Darwin, basing his work in part on the fossil record, contended that higher forms of life emerged from lower forms, with the direction of that evolution determined by "natural selection." In essence, the modern synthesis of Darwinian theory contends that small mutations in individual organisms provided enhanced survival or reproductive value and that therefore the carriers of these favorable mutations were found in greater abundance in future generations. It was Darwin's belief that the accumulated weight of these minute changes, each providing a specific selection value, led to the development of distinct species.[1]

It is this theory of Darwinian natural selection that is now under attack. Jeremy Rifkin and Nicanor Perlas, in their book *Algeny*, state that "the evidence against the neo-Darwinian synthesis is now so utterly overwhelming that it is astonishing to realize that the theory is still faithfully adhered to and vigorously defended within many sectors of the scientific establishment." The contention is that Darwinian evolution represents a reflection of the social and political thought of mid-nineteenth-century England rather than a true scientific principle. Harvard's Stephen Jay Gould, a supporter of Darwin, similarly is reported to feel that at least the part of Darwinian theory that is tied to the *gradual* formation of new species by the accumulation of mutations is a product of Western philosophical bias.[2]

Are we then to believe that higher forms did not evolve from lower forms? Is life continuously being generated about us with still higher forms to be expected imminently? Does selection play no role in biology? In order to answer these questions, we must look at the objections to Darwinian selection and determine what is valid in the theory and what is suspect.

First, it must be pointed out that Darwin himself realized that his theory had not been proved conclusively. One problem that he saw clearly was the fact that the theory required the gradual transition of one species into another. While some cases of such transition have been found in the fossil record, they are remarkably few. Darwin's argument was that at the time of his writing sufficient study of paleontology had not taken place to result in the discovery of the intermediary forms between the various species. Now, over a century later, such intermediary forms have not yet been found and one can justifiably question if they ever existed. It appears from the fossil record more likely that in many cases individual species evolved abruptly from their predecessors, rather than gradually as Darwin suggested. This is the position espoused by Stephen Jay Gould and Niles Eldredge. The potential causes of rapid speciation are the subject of scientific hypothesis and indeed represent one area in which Bioburst technology, as we shall later see, potentially can shed light on the process of evolution.

As early as 1932, the renowned biologist J. B. S. Haldane, among others, discussed the notion of gradual selection working on random variation and found it wanting. He did not doubt that selection occurs within a species. This fact is so well proven as to be self-evident. Antibiotic treatment selects for antibiotic-resistant organisms. The industrial revolution in London selected for the survival of dark-colored moths, which survived better than pale moths in the smoky environment of the city. The deliberate breeding by scientists (a form of "unnatural" selection?) of hypertensive rats can produce an inherited hypertensive tendency. Similarly, thoroughbred race horses are bred for the production of swift-running offspring. There is no doubt that selection can alter the characteristics of cohorts within a species, but from the very first it was clear that sexual selection (that is, selection of traits based on their capacity to increase successful courtship and mating activity) was, when taken alone, inadequate as an explanation for the generation of new species.

Today, the concept of natural selection *in toto* is coming under similar challenge. As Haldane wrote, "selection alone may produce considerable changes in a highly mixed population. A selector of sufficient knowledge and power might perhaps attain from the genes at present available in the human species a race combining the average intellect equal to that of Shakespeare with the stature of Carnera. But he could not produce a race of angels. For the moral character or for the wings, he would have to await or produce suitable mutations." Haldane is not rejecting the concept of evolution. He believes that the evidence strongly supports such an idea, and he explicitly laments the fact that for years after Darwin legitimate attacks on the scientific basis of the theory were obscured by "jabbers of ecclesiastical bombinations." If there is a flaw in the idea that random mutation and natural selection produce new species, it does not necessarily imply that creationism or the theory of the spontaneous creation of life are necessarily correct. It must be recalled that Lamarck himself espoused "evolution," but he believed that the acquisition of new characteristics by the organism was induced by environment, whereas Darwin proposed the idea of selective pressure as the directing force. Nonetheless, all these men, including many of the current critics of Darwin, recognize that evolution is infinitely more likely the correct explanation for the current state of life on earth than is spontaneous creation or creationism. Where they differ with the modern synthesis of Darwinian theory is in the notion that small random point mutations coupled with natural selection form the sole mechanism by which new species are produced. Indeed, statistical arguments can be given to the effect that, on the basis of random point mutations in structural genes, higher forms are extraordinarily unlikely ever to evolve. Whether point mutations in regulatory or "control" genes could suffice remains problematic.[3]

The preponderance of evidence and opinion suggests that species developed by evolution, and that selection is a valid principle for explaining many, but not all, of the different characteristics of various cohorts within a species (some traits become fixed within a species not because of selection but rather more or less randomly through the process of "genetic drift"); but that random point mutations, even coupled with natural selection, may be inadequate to account for the observed course of evolution. Haldane himself concludes that "the actual steps by which individuals come to differ from their parents are due to causes other than selection and in consequence evolution can only follow certain paths. These paths are determined by factors which we can only very dimly conjecture."[4] Although he was unable to explain the driving force behind evolution, Haldane saw merit in a genetic theory of evolution in which the changes between species developed from the laws of population genetics and natural selection, coupled with unknown processes generating biological variation, rather than solely from random mutations and selection.

This view becomes more pertinent when one considers the dynamic nature of the gene as elucidated recently by molecular biology. DNA transposition, message processing, and other gene activities provide an enormous complexity against which new permutations can potentially emerge. We can envisage fluctuations in gene number, location, and action producing speciation much more abruptly than random point mutations ever could. Additionally, modern techniques of hybridization analysis potentially shed light on the genetic interrelations of species and perhaps may provide the data that will lead to a clarification of evolutionary theory. The basic principles of evolution and selection have not, after all, been invalidated in the current debate; rather, their precise role in the origin of species and the genetic mechanisms by which they operate are being questioned. Bioburst technology can potentially provide definitive answers.[5]

For an example of how genetic mechanisms can lead to directed, as opposed to random, change, one can look to the concept of "molecular drive," as reviewed by Gabriel Dover of Cambridge University. Dover explains why evolution need not progress solely by minute changes, the sum of which produces a new species. Rather, genetic events such as DNA transposition may be capable of producing changes in organ function and structure in a discontinuous fashion as opposed to the gradual transitions predicted by the classical theory of natural selection. But this possibility does not discredit natural selection, for selection still participates in the determination of which alterations will survive.[6]

Theories of "molecular drive" and the like show that the detailed study of evolutionary mechanisms on the level of molecular genetics is potentially a tremendously fruitful field. Additionally, homology studies between the genomes of various species are daily gaining importance in the effort to define the precise evolutionary relationship between species. New and unsuspected relationships between species are being detected on the basis of genomic DNA analysis. The prospect for advances in this area seems bright.[7]

Molecular biology will likely have a great deal to say about the specifics of evolution, but one thing seems clear—Darwinian natural selection is not dead. It has simply, albeit abruptly, evolved to a higher form.

CHAPTER XVII

Bioburst and
the Philosopher

Although the major concerns of organized religion regarding molecular biology seem to center currently on disruption of the human germ line, there are additional issues that may also face the theologian-philosopher. For example, do the new insights provided by molecular biology teach us anything about consciousness or the nature of man?

Consider, for example, the problem of defining the humanity, or lack thereof, of a living entity. This is the underlying conundrum in the current abortion controversy, and it will likely re-emerge in many guises as future Bioburst advances are employed in the area of human reproduction. Just as we found in chapter 1 that *life* is difficult to define precisely, we now find it difficult to define *human* life precisely. And just as we found it necessary to introduce a somewhat arbitrary definition of life, we now encounter similar difficulties regarding the definition of human life. On the one hand, there are those who ascribe humanity to a living organism based on the nature of its genetic composition, arguing that all the designs and plans for a new human being are contained in its unique genome, irrespective of whether that genome has already done its work by producing a grown man or resides in the potential-laden environment of a single fertilized germ cell. The *potential* for human life is considered sufficient. On the other hand, there are those who believe that the informational content of DNA is *contextual*—that is, the genetic information of the fertilized ovum must be translated into RNA and protein and cells; the cells must rearrange themselves into organs, and the organs into a human being before the information becomes palpably significant. In this view, the point at which the contextual information of the new being reaches the level of "humanity" remains arbitrary. Which of these positions one chooses to adopt depends more on one's philosophical or theological views than on biological insights. The precepts of molecular biology can make more vivid the arguments of each side, but they do not provide much insight into this important, and quintessentially human, problem. The issue is at heart a matter of values, not of biology.

A variation on this basic theme is the current controversy surrounding the "case of the frozen embryos." In this instance, a healthy infertile couple desirous of children arranged to undergo *in vitro* fertilization. As is usual, several ova were harvested from the woman and fertilized with sperm (there is some question in the original case as to whether the sperm was actually that of the husband or of an unrelated donor, but this complicating factor will be ignored in the following analysis). After several cell divisions, one embryo was implanted into the woman and the others frozen in case they were required by virtue of failure of the initial attempt. As it happened, the pregnancy was not viable and miscarriage occurred. Rather than immediately re-attempting the procedure, the couple elected to seek some psychological respite from the stress of their undertaking by taking a brief vacation. Unfortunately, they were killed in an airplane crash. This tragedy posed several problems regarding the fate of the remaining frozen embryos. If these are to be regarded as independent human beings, they must at least be maintained alive in their frozen state and ideally implanted into a willing surrogate mother so that they might be given a chance for normal growth and life. On the other hand, if these embryos are viewed as simple multicellular organisms they presumably belong to the couple's next of kin to do with as they choose—including permitting them to thaw and die. Even more complicated is the question of inheritance. If each embryo is deemed to be a human being, it presumably has some claim to inheriting a portion of the fortune of its dead parents. If considered a primitive organism, it cannot even command sufficient funds to pay for the electricity that keeps it in its frozen limbo.[1]

Although the case just described may seem uncommon because of the death of the parents, it must be pointed out that, for practical reasons relating to the low success rate of the *in vitro* fertilization procedure, multiple embryos must be produced in each case where a successful pregnancy is attempted. What right to life do they have? What claims can they make on society? Answering these questions on a humanistic as opposed to a theological plane requires that a basic decision be made regarding the intrinsic value of the informational content of DNA in a germ cell as opposed to the contextual information generated during the course of development. This is fundamentally a philosophical issue of values, not one of molecular biology.

This is also the case with human consciousness. The mystery of conscious self-awareness has fascinated man for centuries. As science learned more and more about the workings of the human machine, it sought in vain for the soul or the mind. The so-called mind/body problem has haunted philosophy, medicine, and biology since human inquiry began. Only theology seems at home with this conundrum. To be sure, scientists like Dr. Roger Sperry have extended man's understanding of consciousness through the detailed study of people who suffer from neurological disorders—such as patients whose brains have been surgically altered in an attempt to control severe seizures. The operation performed on these people consisted of severing the corpus callosum and other nerve tissue connecting the right and left hemispheres of the human brain, and it was expected that the procedure would have the effect of limiting seizure activity to one hemisphere or the other. However, such patients then behaved as if they possessed two compatible, though distinct, brains. Because in most people the left hemisphere contains the speech area, only this hemisphere can verbally interact with the outside world. And yet when the written names of objects are presented to the right hemisphere—by flashing them in the left visual field—the right hemisphere is capable of reading the written word and using the hand under its control, the left, to pick the objects out of a group—all without the left hemisphere knowing anything about it. At times, minor differences can develop between the two hemispheres, so that, for example, one hand attempts to take off the pants leg the other just tried to put on. Upon learning of these results, one feels compelled to inquire about the nature of the consciousness of the mute hemisphere. Can one sense its presence in the normally connected brain? Should the first-person singular pronoun be abolished? Is every "I" really a "we"? And if the right hemisphere is conscious in so apparently inconspicuous or deferential a fashion in normal man, should one consider the possibility that lower-order neural structures such as the spinal cord—or for that matter lower organisms such as a snail—possess a consciousness of their own? This exciting work has raised more questions than it has answered.

Does molecular biology have anything to tell us about this matter? Yes—and no. Studies of brain neurons, coupled with modern techniques for the detection of small quantities of hormones and peptides, have revealed a startlingly complex picture of the human brain. Instead of only the few simple chemical signals that were known to function as neurotransmitters twenty years ago, it now appears that the brain contains and probably uses "to think" a large array of small proteins previously thought to function solely as hormones in the peripheral circulation or in the gut. Angiotensin II, cholecystokinin, thyrotropin-releasing hormone, beta-endorphin, enkephalin, and others have been found in the brain, greatly expanding the potential complexity of the nervous system.

Additional studies seem to suggest that, in the brain, the nerve cell (*neuron*), instead of acting like a simple adding machine that sums up the input of the inhibitory and excitatory neurons impinging upon it and then decides to generate a neurosignal based on whether the sum is positive or negative, actually works like a cellular computer chip. The location of inhibitory and stimulatory inputs on the nerve cell surface may be important, leading to a built-in logic in the cell. If this is true, the nervous system is far more complex than previously believed. Nerve cells consist of a central body, where the nucleus and cell machinery are located, as well as of antennalike extensions called *axons,* which join nerve cells with adjacent nerve cells and form the wiring of the brain. If the hypothesis is correct that inhibitory inputs close to the cell body inhibit *all* positive inputs distal to them along the axon, then these inhibitory neu-

175

rons form a "but-not" logic circuit. Consider, for example, the neuron shown in figure 25. We see inpinging on this neuron the axons of six other neurons, three of which are inhibitory, marked with a negative sign, and three of which are stimulatory, marked with a positive sign. If all of these input neurons were to fire, we would have three positive and three negative inputs, and by the old theory of summation the target neuron would see no effective input and would not fire. However, if the more complex model, implying as it does that inhibitory neurons block the effect of all stimulatory neurons distal to them, is correct, then it can be seen that, for example, inhibitory axon D would block the stimulatory input of axons C and A. Thus, the logic circuit formed by A, C, and D (that is, assuming that these were the only inputs to fire) would read: target neuron to fire if input comes from A or C "but not" D. And, in the schema outlined, should all inputs to the neuron be activated, the target neuron would fire because axon E would provide a stimulus close to the cell body that is not blocked by any negative input. It can be appreciated, then, that if this theory of neuronal action is confirmed, another level of complexity has been uncovered in the logic circuits of the brain. Many more undoubtedly exist.[2]

Recombinant DNA technology is currently being applied to the study of the genes controlling behavioral responses in mollusks such as aplesia, and so may determine the genetic underpinnings of neuronal wiring patterns and instinctual behavior. Presumably, abnormalities in gene expression in these primitive animals could lead to faulty brain transmitters and aberrant nervous system behavior. Studies of this genetically produced behavior likely will shed light on the normal functioning of the nervous system of these simple animals, perhaps thereby providing insights into the nature of human intelligence. Additionally, molecular biology is investigating the possibility that regulation of the receptors for transmitter substances located on neurons could play a role in such processes as memory and thought. Even more intriguing is the possibility that genetic plasticity could play a role, with alterations in gene elements affecting receptor or transmitter levels, so as to produce the basis of memory. Already it has been suggested that genes are transcribed in an odd way in the brain. First, RNA polymerase III transcribes a

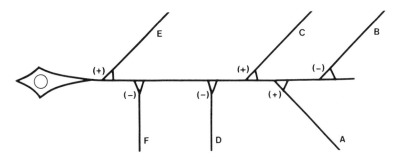

Fig. 25. Logic circuit of a neuron.

portion of the gene, and only then does message-related RNA poly-merase II generate messenger RNA in the usual fashion. The significance, if any, of this unexpected mode of transcription for memory or the ability to think is unclear at present, but it seems an intriguing clue. Clearly, major breakthroughs in science's ability to describe and analyze the cellular and molecular basis of thought are likely in the near future.[3]

Despite all this potential for elucidating the complexity of the nervous system, science is still no closer to an understanding of self-awareness. Douglas Hofstadter, in his book *Gödel, Escher, Bach,* hypothesizes that, with the increasing complexity of logic networks, a sense of self-awareness can develop in a machine or an animal. Consciousness resides not in one cell or wire but rather in the logic network. Moreover, such "higher order" systems could well be impenetrable to the analysis of the "lower order" systems that comprise them or even to the higher order systems themselves. Only a still more complex system could fathom them. Hence, the inexplicability of consciousness. This or similar views are championed by many students of artificial intelligence and computer science. Additionally, many of these scientists go further and contend that the development of artificial intelligence is potentially close at hand—and with it perhaps will come a greater understanding of human intelligence.[4]

In arriving at these conclusions, computer scientists fall back on the so-called "Turing test" for determining intelligence in a machine, in large part because there appears to be no other readily available or applicable criteria of prehension. The basis of the test is simple, although this simplicity belies the complexity implicit in successfully passing it. The Turing test requires that a machine being tested for artificial intelligence be placed in a room and connected by teletype to two men located in two other separate rooms. The two men are also connected by teletype lines. One of the men then initiates discussion by teletype with either the machine or the man in the other room, without being told which is which. By teletype he can interrogate the occupant of Room 1 and the occupant of Room 2 in any way he wishes and, in turn, he receives their responses by teletype. If, after a period of interrogation and "conversation," he is unable to distinguish the man from the machine, then the Turing test criteria are deemed to have been met and the machine is said to possess artificial intelligence. Presumably because men are intelligent and the machine behaves indistinguishably from a man, the machine possesses intelligence.

Although a logic system of sufficient complexity could perhaps duplicate the responses of the human nervous system in the face of this kind of testing and thus possibly possess "artificial intelligence," there is nothing in the Turing test that compels the observer to assume that this artificial intelligence encompasses what we term "consciousness." Hofstadter suggests that, as logic systems become more complex, feeding back into themselves and developing subroutines that monitor or analyze the systems themselves, consciousness results as a natural consequence. Clearly, this view admits the possible acquisition of consciousness by machines possessing artificial intelligence, yet this consciousness may be incomprehensible to both man and machine. For example, one could perhaps argue that the famous theorem on indeterminate proposition of the logician Kurt Gödel—demonstrating that mathematical logic systems cannot be both complete (by *complete* is meant the ability to derive all true theorems of the system using only the axioms and rules of the system itself) and consistent—can be extrapolated to the philosophical realm by saying that the "part can never entirely comprehend the whole." Just as Gödel's proof tells us that the totality of all arithmetic truths cannot be derived from the axioms and rules of arithmetic, it might also suggest that the nature or underpinnings of consciousness cannot be fully deduced from the functioning of consciousness itself. To put this another way, does not Gödel's proof tell us that any system of logic *must* be blind to the totality of its nature and implications, not to mention its axiomatic origins? If the blindspot that this Gödelian "unknowability" produces in our understanding of the mind should concern an important facet of consciousness, then the nature of consciousness must in part be logically "unknowable" as well. In this case, although consciousness could be a fundamental physical property of circuits, it must always remain an entity that is not totally comprehended.

The idea that consciousness is unknowable (in terms of current laws of physics and chemistry) has actually been subjected to fairly detailed, albeit controversial, analysis by quantum physicists over the last fifty or so years, although the nature of physics rather than Gödel's theorem served as the impetus to these investigations. For example, Erwin Schrödinger himself wondered about man's current capacity to understand life, and W. M. Elsasser has taken the position that it is possible that life is *in principle* too complex to be understood even in terms of quantum mechanical laws. Life should, he argued, perhaps more properly be defined in terms of "biotonic" laws distinct from quantum physics—a premise that represents a considerable departure from conventional twentieth-century thought and that was enunciated prior to recent breakthroughs in the understanding of molecular biology. The renowned physicist Eugene Wigner took a similar, though less inclusive, position and suggested that consciousness (as opposed to life in general) had best be viewed as directed by biotonic rather than quantum laws.

More specifically, consciousness in this view interacts with physical reality in a way different from that which characterizes the quantum mechanical interactions of unconscious bodies. These mind-matter interactions will only be defined by yet-to-be-discovered physical laws. In addition, Wigner argued that the ability of the conscious mind to make unambiguous observations about reality, as opposed to the probabilistic predictions of quantum theory, suggests that the interaction of the conscious and the unconscious is best described by certain nonlinear equations rather than by the linear superimpositions commonly used to describe the inanimate world. To be

179

sure, this view is not widely accepted, and the quantum physics of thought remains under active study today. Indeed, there are those who feel that matter and consciousness represent different manifestations of the same higher reality, and these scientists are currently attempting to formulate equations capable of describing this alleged state of affairs; others see parallels between modern quantum mechanics and the mystical view of mind and universe espoused by Eastern philosophies. Nonetheless, all these hypotheses at least serve to point up the controversy and difficulty associated with attempts to describe consciousness in physical terms—which in turn raises the possibility that consciousness may not only be less susceptible to the analysis of man than is unconscious matter, but that if explicable in terms of present-day physics and physiology, it is so only in an imperfect sense.[5]

In any event, molecular biology does not seem to add much to the philosophical aspect of this controversy. It does add tremendously to our understanding of the wiring and circuit logic of the brain, however, and it offers the potential for therapy and constructive intervention. But it has not, and almost certainly cannot, clarify the basic philosophical dilemma. These issues dealing with the nature of life and consciousness demonstrate that for all its power Bioburst cannot answer all questions or explain all biological phenomena. Molecular biology is an incisive, incredibly useful branch of science, but it is not omnipotent.

CHAPTER XVIII

Bioburst and Management

A chemistry professor once remarked to his students that had Pasteur, when he discovered the unusual and previously undescribed levorotatory (that is, possessing the ability to rotate polarized light to the left) properties of certain crystals of tartaric acid, simply washed the offending crystals down the drain as inconsequentially aberrant, the entire concept of stereoisomerism would have been missed. Chance favors the prepared mind, and this is nowhere more true than in the arcane world of molecular biology. The need for a staff well versed in all areas of biology has always existed in biomedical research, but the need is all the more acute now that so apparently great an intellectual gulf separates the old and the new biologies. That this schism is real can be readily seen by discussing the issue with the faculty of any medical school. The large majority of these men are not involved in the Bioburst phenomenon and in many ways resent those who are. They feel that the new breed is elitist, devoted to the study of very narrow biochemical issues, and is moreover the undeserving recipient of an unreasonable amount of grant support. The new breed, for its part, too often retreats into its own world. This situation is unfortunate and unnecessary, and it will be counterproductive if permitted to persist. There are some who wish to bridge the gap, but clearly a new breed of scientist capable of interfacing with both camps must spring up if a synthesis of the insights of both groups is to be achieved and the productivity of each is to be maximized.

If it is hard to predict what form this interfacing will take in academia, it is even more difficult to speculate about the response of industry. Yet the successful meshing of the molecular biology of Bioburst and the tenets of classical pharmacology is absolutely critical to the very survival of the domestic pharmaceutical industry. It is apparent to many that the American drug industry is rapidly losing its innovative edge to foreign concerns. Failure to lead in the expanding application of Bioburst technology to the design of new pharmacological agents could actually spell the end of the domestic industry. This fact is not lost on the industry's foreign competition. Conceptually, the worst response of a traditionally organized pharmaceutical firm to this challenge would be a rigid adherence to standard pyramidal structures in management and research. The inflexibility inherent in this approach would almost certainly impede the efficient exchange of ideas and information between modern molecular biologists and traditional pharmacologists. The new molecular biology is pluripotential and must interact with many disciplines to reach its full productivity. The pharmaceutical industry has always been aware of the value of research cross-fertilization and even of serendipity, but the new era will place even more emphasis on, and test more severely, the flexibility of this industry than anything that has gone before.[1]

A recent example illustrates this point. Sandoz, like many other pharmaceutical firms, asks its employees to collect, whenever traveling abroad, soil samples from the areas they visit. Dr. Jean Borel was the man at Sandoz responsible for evaluating the products of fungi found in these samples for their potential efficacy as chemotherapeutic agents. When a new fungus was isolated from samples obtained from Norway and the United States, Sandoz scientists evaluated a water-soluble substance that it produced for possible antibiotic activity. They found little. Based on his past experience regarding the interrelation, or lack thereof, of antibiotic and immunoregulatory properties, Borel elected to check this material, now known as cyclosporin A, in assays designed to detect immunological activity. To his surprise, the compound suppressed T-cell-mediated immunity, but in contradistinction to other suppressors of cell-mediated immunity, cyclosporin A did not kill T-cells. Rather, it simply arrested their proliferation in response to antigen. Thus, the T-cell effect was reversible, and the drug's toxicity was, therefore, substantially less than that of previously used agents. Company management was not initially enthused by the project and wished to cancel it as part of a general reassessment of immunology at Sandoz, but Borel persisted and eventually was persuasive.

In recent clinical trials cyclosporin has proven to be the most useful suppressant of transplant rejection yet discovered. It has revolutionized transplant technology to the point where liver transplantation and similar procedures are arguably routine rather than experimental. Indeed, the availability of this drug may so dramatically change the transplantation field that shortages of organs for transplantation will become critical. In addition, the drug may prove useful in the therapy of autoimmune disorders and may even prevent the development of diabetes in some susceptible patients. Although the financial payoff to Sandoz is still unclear, it is likely to be substantial.

In analyzing this success story, several points should be noted. First, it was the free flow of knowledge, ideas, and hunches between microbiologists, immunologists, and management that saved the project. It is a credit to Sandoz that it did not permit its plans for the redirection of immunology research to prevent continued development of cyclosporin A. But it is most of all a credit to Borel's persistence and, in the final analysis, to a management that eventually listened to him.

No one knows exactly how cyclosporin A works. Its end result—T-cell suppression—is appreciated, but not its mechanism of action. For the purpose of argument (and purely hypothetically), suppose that this drug is taken up by cells and alters DNA structure by converting stretches of the double helix into the so-called Z-form, and further assume that this action somehow accounts for its effects on T-cells by preventing the production of T-cell growth factors.[2] Then suppose that this action on DNA was known, but the final effect on T-cell-mediated immunity was not. How much harder would it have been to convince management to keep the project alive? Many of the great insights of the future are likely to come in this way—as biochemical observations related to DNA gyrase, or polymerase, or what-have-you. The insights will arise first in a

183

recombinant DNA laboratory and not in a Norwegian field, and the ultimate implications and consequences of the application of these insights will not be immediately obvious at the time of their discovery.

This is a radical departure from the way pharmaceutical advances occurred in the past. Heretofore, assays of biological potency were used to screen a vast number of compounds, with efficacious products identified by this random selection process. Thus, many pharmaceuticals were developed by a mechanism much akin to nature's developmental scheme, natural selection. At other times, chemists produced analogs of compounds known to be biologically active so as to enhance one or another aspect of the compounds' actions. These processes—the screening of organic compounds for biological activity and the chemical synthesis of analogs to active compounds—have accounted for the vast majority of products of the pharmaceutical industry to date. But breakthroughs in the future will come from the drawing boards of the molecular engineer, and, although rich in potential, they will arrive in the world in a relatively obscure form. Management and scientists alike will require a new way of looking at things in order to realize the full importance of such observations and to assure that these findings are transferred to the immunologists, microbiologists, and others who can put them to work. New working relationships between product-seeking management teams and knowledge-seeking academic teams must be developed. This is the challenge of Bioburst to management. And this challenge can be met and the needed insights developed only if management can couple native wit and creativity with a sound grounding in biology.[3]

Solid education in biology, molecular and otherwise, will be indispensable to the Bioburst industry employee, be he or she a technician or a company president. Students interested in preparing to work in this industry must plan their study accordingly, and schools must respond with courses to meet their needs. This educational process could profitably begin in junior high school. Pharmaceutical managers who have not been schooled in molecular biology should undertake the comprehensive study of the subject by way of the three traditional means open to them: learning from newly recruited scientists, interaction with consultants, and self-teaching. Business schools would be wise to join with their basic science colleagues in offering programs of study designed to aid in this knowledge acquisition by senior management.

Because molecular biology is so basic, its ramifications are quintessentially multidisciplinary. Management structure should reflect this characteristic and should be as fluid as possible so as to provide maximum input from all parties and to assure the sharing of information across departmental or division lines. Even then, new and creative modes of interacting freely with academia must be developed. Only in this way will productivity be maximized.

CHAPTER XIX

Bioburst and
Life's Buffers

These essays have touched on the nature of life, reproduction, genetics, and even consciousness. They have described how modern molecular biology can provide new insights into the basic fabric of life and at the same time produce technical advances of enormous importance. We have seen how rare drugs can be synthesized in great abundance, and how new techniques can detect abnormalities in the genes of patients, born or unborn, who suffer from heritable diseases of metabolism. We have seen how Bioburst has provided new insights into the nature of cancer, heart disease, and stroke, and how it has offered the potential not only for the development of effective therapy for these diseases, but for their ultimate elimination. Modern molecular biology may provide us with vaccines against cancer, crops that grow in arid locales, and the ability to lessen or entirely eliminate most organic pollution from our environment. Moreover, Bioburst offers the possibility of attacking the aging process and, if carried to its extreme, the potential to produce biological immortality. We have seen how Bioburst can make us healthier, better fed, longer lived, and even—either directly or indirectly—happier.

We have also considered the intellectual ramifications of molecular biology, including its impact on the ways we view ourselves, educate ourselves, and conduct our affairs. The host of ethical, moral, and social questions raised by this technology cries out for solution.[1]

How are we to approach Bioburst on the basic, personal level?

How is it to be blended into the mixture of experiences, insights, and teachings from which we construct our world? How are we to buffer our beliefs from its impact? These questions are not trivial, dealing as they do with the self-consciousness not only of individuals but of entire societies. Bioburst must be amalgamated with the experiential past of man in an acceptable, insightful, and productive way.

The philosopher Martin Heidegger alluded to questions much like these in his well-known Memorial Address presented on the occasion of the 175th birthday of the German composer Konradin Kreutzer. Discussing this issue of science and man, Heidegger wrote:

> The international meeting of Nobel prize winners took place again in the summer of this year of 1955 in Lindau. There the American chemist Stanley had this to say—"The hour is near when life will be placed in the hands of the chemist who will be able to synthesize, split and change living substances at will." We take notice of such a statement, we even marvel at the daring of the scientific research, without thinking about it. We do not stop to consider that an attack with technological means is being prepared upon the life and nature of man compared with which the explosion of the hydrogen bomb means little. For precisely if the hydrogen bombs do *not* explode and human life is preserved, an uncanny change in the world moves upon us.
>
> Yet it is not that the world is becoming entirely technical which is really uncanny. Far more uncanny is our being unprepared for this transformation, our inability to confront meditatively what is really dawning in this age.
>
> No single man, no group of men, no commission of prominent statesmen, scientists, technicians, no conference of leaders of cumbersome industry can break or direct the progress of history in the atomic age. No merely human organization is capable of gaining dominion over it.
>
> Is man, then, a defenseless and perplexed victim at the mercy of the irresistible superior power of technology? He would be if man today abandons any intention to pit meditative thinking decisively against merely calculative thinking. But once meditative thinking awakens, it must be at work unceasingly and on every occasion.[2]

Heidegger's words can be understood on two levels. On the first, he provides a prescription for dealing with Bioburst, atomic energy,

187

and the new scientific revolutions that are yet to come. Although he offers no specific solution, his prescription is inherently sound in its approach. Meditative thinking—the unfettered contemplation in depth of man's roots and the mystery inherent in his existence, as viewed against the background of the past, the present, and the likely future—by both the individual and society offers the prospect of gracefully molding Bioburst into the contextual fabric of man's history. This course requires neither the abandonment of technology nor the denigration of man. Rather, it offers the possibility of blending man and science rather than pitting one against the other. In the absence of meditative thinking, disquiet, malaise, anxiety, and "future shock" are inevitable. If meditative thinking is practiced by the individual and, to the extent possible, by the collective, the splendid achievements of the new biology and their enormous promise will be seen in their true beauty as a natural part of man's intellectual odyssey. The needs and fears of man will temper but not thwart scientific advance, while science attempts to fill man's needs and assuage his fears. This is Heidegger's message and the challenge to us all.

On a higher level, Heidegger intended his remarks in 1955 to be an analogy for the proper approach to understanding man's "being" or "consciousness." In his philosophical writings, he argues that through meditative, as opposed to calculative, thinking, insights into the intrinsic nature of "being" can be glimpsed—and in fact only through this meditative modality can these insights be revealed. In dealing with issues like life, aging, and most particularly consciousness, modern science may be trying, with the tools of calculative thinking, such as molecular biology, to explore concepts that require meditative thinking for their understanding. If so, Heidegger would argue, we are doomed to failure. Only taken with something more, something richer, can the technical lessons of Bioburst shed light on the true nature of life or explain, if that is possible, such conundrums as the mind-body "dichotomy." Of course, an explanation of the mind or consciousness or of a thousand other phenomena may forever elude us. We may come no closer to it perhaps than obtaining an inclination or impression of what the truth might be.

Along these lines there is much to be learned from the approach to consciousness taken by Douglas Hofstadter in *Gödel, Escher, Bach*. Hofstadter's analysis, as we have seen, argues by analogy with Gödel that consciousness (and therefore perhaps "being" or "soul") may be a material but unknowable process. He suggests that the higher-order workings of the nervous system cannot be fully known by the subordinate portions of the system. Gödel, it will be recalled, demonstrated mathematically that no reasonably complex system of logic can at once be consistent and entirely self-contained. No system can provide proof of all its theorems and remain consistent. By analogy, the nature of thought may not be fully knowable to the process of thought itself. This argument is extrapolated by Hofstadter to the view that consciousness may grow out of a logic system so complex that the system develops a rich symbolic representation of itself that it then feeds back into its own workings. Self-awareness and consciousness are the result. This self-awareness is a material consequence of the circuitry of the brain. But, by reasoning analogous to Gödel, the nature of the ultimate process—its workings, if you will—may be in part intrinsically unknowable to the brain itself (with all apologies to the sensitivities of logicians and the beauty of Gödel's proof). Although consciousness is rooted in the materialism of biology, any attempt to fathom its nature in a scientific sense must ultimately be based on the disciplines of mathematical logic and microscopic physics. Logicians tell us that our mathematical logic possesses a disturbing incompleteness that suggests it will fail if pressed to describe the higher-order logic system that is consciousness. Scientists tell us that our physics may be incapable of providing, at either the cellular or molecular level, complete descriptions of the substrate of consciousness. All of this suggests that while, for example, the workings of the heart may yield under the reductionist onslaught to a degree that satisfies our conscious minds, the nature of consciousness will not. One can thus view consciousness as material but unknowable, something that we can only glimpse in a shadowy, misty way through the contemplation of higher-order logic groups. Thus, research into artificial intelligence and the theory of knowledge seems to lead to the same world that Heidegger's meditative thinking beckons us to. Perhaps this is as close to the ultimate as we shall ever come.[3]

The theologian, of course, could agree with these premises, accept fully the idea that consciousness is an unknowable process, but contend, in contradistinction to Hofstadter, that it is at heart immaterial (*i.e.*, spiritual). After all, if consciousness is unknowable, its nature, be it material or immaterial, is necessarily unknowable as well. There is no ultimate truth or conclusion in these Gödelian analyses that preferentially favor the views of the believer or the disbeliever. Perhaps meditative thinking of an even more intense and personal sort can serve to make these distinctions clear for the individual. It cannot for the logician.

But what if this concept of the unknowable is extended from an analysis of consciousness to the analysis of life itself, or of "life ecology"? The existence of life as an entity requires the careful balancing of forces and the careful integration of processes spanning the spectrum from the quantum mechanical to the species level. Life is a process of ultimate complexity involving interactions with a physical environment, a whole host of living organisms, and blind chance. The ultimate nature of this process may not be knowable or understandable by any of its parts, including man. Thus, the ultimate implications of manipulating life also may be unknowable or incomprehensible to man. In dealing with this unknowability lies a great challenge that must be directly faced.[4]

Throughout this discussion of Bioburst, we have compared the risks and benefits of this new technology to the risks and benefits associated with more standard biological manipulations that are accepted as part of everyday life. Might it now be argued that, although recombinant DNA technology is different from, say, selective animal breeding or human artificial insemination, it is different only on a technical rather than a theoretical level? Indeed, this argument appears to be valid for those cases that we have analyzed in detail. However, we must also recall that there are instances, such as the deliberate freeing of engineered organisms or the widespread use of germ line modification, in which we could not be so sanguine about ultimate safety. Caution, contemplation, and oversight are urged in these areas, not because the projects proposed to date are intrinsically harmful or qualitatively different from activities that have gone before, but because their consequences may in a real sense be "more unknowable" or less predictable than the consequences of previous biological manipulations.

Life is a "buffered system in equilibrium." Just as a chemical buffer resists any attempt to change its pH, life forces, from the genetic to the ecological, resist any change from the equilibrium of survival. If inadvertently kudzu is unleashed upon the South, or killer bees are bred in Central America, natural forces will eventually control them and bend them to the basic theme of life on the planet. Life in this sense is forgiving of the bungling attempts of man and nature—at least so long as the magnitude, or dose, of a distorting change is not enormous.

But changes on the most basic level—the level of quantum chemistry and molecular genetics—mean that this buffering is much less, the potential swings from the equilibrium much greater, the potential consequences of minor alterations more severe, because the buffering and forgiving and repairing capacities of the genome itself have been tampered with. The system is potentially less forgiving and the consequences of error more severe. For this reason, the dose of such alterations must be more closely regulated, for much of life's capacity to buffer changes is bypassed in the process of genetic engineering, leaving ecology as the prime buffering agent. We must always remember that the net consequences of any unbuffered disequilibria for the higher-level-system resonances that are the essence of ecological balance, and indeed of life, are unknowable.

For the first time, science is working directly and frequently at the basic level of life. It is now entering the world of quantum chemistry and the theory of knowledge, knocking up against the unknowable. It is entering a less-buffered world. This situation is quantitatively, but not qualitatively, different from that faced by Cotton Mather, whose small-scale deliberate spreading of smallpox virus also involved risk, at least to the people of colonial Boston. This quantitative difference is important because it implies that the margin for error today is less, and the need for oversight and regulation is greater. It does not imply that the new science is evil or that its benefits must be shunned, any more than variolation should have been prohibited in early Boston. Rather, it argues for moderation in the application of the new technology.

On the practical level, there is great promise in the technology of modern molecular biology, with only minor short-term risk. But in the long term there are definite, albeit subtle, risks associated with such activities as free-release experiments and large-scale germ line therapy. The same cautious concern that a physician or midwife shows for the ultimate spiritual, physical, and mental well-being of a woman in childbirth must be shown by society for life as an entity and more particularly for the emerging life of future generations. Decisions regarding bioengineering must be made with great empathy for life and the environment. Society must learn from the mistakes of the past, from the toxic waste dumps and polluted drinking water that plague us today as a result of popular disinterest years ago. The errors of ignorance and indifference that led to the Love Canals of today must not be permitted to produce still worse disasters tomorrow. Policy must be widely discussed and based on broad consensus. The issues cannot be ducked or left to technical experts, because the policies adopted will depend as much on philosophy as on science. Practical questions must be asked and answered dispassionately, and unknowables must be addressed directly. The unknowable consequences of science's proposed actions must be squarely admitted, as must the unknowable (and knowable) results of allowing "nature to take its course." Life's multiple buffers must be appreciated and respected. Man's psychic buffers must be protected and called into play to protect his spirit. Moderation must temper recklessness without stifling exuberance. Individually and societally, we must all practice meditative thinking of the highest sort if the full fruits of Bioburst are to be safely harvested and its errors harmlessly blunted.

CHAPTER XX

Fantasy-Theory: Super Zen

At the heart of these essays are found notions of life, consciousness, knowability, and the proper use of knowledge. These concepts, it must be admitted, are in many ways extremely Western in their origin. The dichotomy between life and death, mind and body, might seem, at best, artificial and irrelevant to the schooled Eastern mind. The oriental view teaches that man and universe are one, variant forms of the same essence, the one not intrinsically different from the other. Life is a valued, beautiful entity, but it is no different in fundamental nature or value from other parts of the cosmos. Personal disaster dissolves into the background totality of all things, with inner peace and acquiescence the result. Indeed, the scientific study of the origin of life, the nature of consciousness, and the arrow of time lend support to some of these ideas.

But what of molecular biology, genetic engineering, and quantum chemistry? Can these ideas, and the worst-case consequences of their application, be blended into Eastern analysis? Can nuclear annihilation? The second law of thermodynamics? Can a certain harmony be seen in the vaporization of the world into space plasma or in the unwinding of the DNA helix by misguided engineering? Probably so. There is a certain tranquillity—perhaps even a Nirvana of sorts—to be found in the gradual thermodynamic winding down of the universe toward equilibrium.[1]

At the same time, the contemplation of these things by Western man carries with it a certain sense of helplessness, of disquiet, of disarray. These feelings are only intensified by the realization that

193

every facet of man's life lies under the sway of the second law of thermodynamics, a cruel master the origins of which are only now beginning to be revealed by the study of cosmology. At present, man cannot with any certainty predict the future of the cosmos in which he lives, much less the future of that segment of the universe known as "life." But he knows that for the foreseeable future he and his kind will be governed by the law of increasing randomness ("entropy") and any attempts to thwart its rule are for naught. In his efforts to explain and influence the world about him he has turned to ever more sophisticated theoretical constructs, with the result that his analytical capacity has been extended—by quantum mechanics—to the realm of the smallest particles and the strongest forces. But still his capacity to predict and control nature remains limited.

Indeed, quantum mechanics has its own problems. It is at heart a probabilistic discipline whose legitimacy has troubled many, including Albert Einstein ("God does not play dice with the universe"). One consequence of the mathematical structure of this science, for example, is that the descriptive equation ("wave equation") of an object includes terms dealing not only with what happened to the object in the past but also with what didn't happen. An analogy might serve to make this point clearer. Consider the wave equation of a soldier just home from the war—accepting for the moment that so enormously complicated a function can actually be written down—and note that this equation contains not only elements derived from all his victories but also terms related to his having been killed by each bullet fired at him. To be sure, the mathematical coefficients of the terms describing the dead soldier equate to a very low probability of outcome, so that when an appropriate calculation is done on the wave equation of the living soldier home from the war, the prediction is a very high probability that he is alive. Yet the inexactitude of this result is troubling, for while considering a live soldier it is difficult to accept that, at least until the observer actually sees the live soldier, physics predicts with some minute probability that the soldier is "really" lying dead on a battlefield. This inability of quantum theory to provide definitive descriptions of reality is one troubling feature of the theory. At heart, the issue is that the quantum mechanical description of a soldier (or more realistically of a beta particle) *must* contain terms related to the soldier having died (or the particle not having been emitted), until the soldier is seen alive (or the particle counted) by an observer, at which time, according to prevailing theory, the wave equation of the observed event "collapses" to that of a live soldier (or an emitted particle). Only through this "collapse" can the predictions of quantum mechanics in the atomic domain be reconciled with the macroscopic world. If the "extraneous" terms characterizing the soldier as dead or the particle as nonemitted are dropped before observation, the wrong results of physical experimentation are predicted.

In addition, the effects of observation are "nonlocal," in that once an observer makes an observation about one part of a system he immediately is entitled to remove a good deal of the ambiguity associated with the observed variable from other parts of the system. For example, electrons possess "spin" of two sorts—up and down. The wave equation of an apparently upward spinning electron, however, does contain a term implying that (with small probability) the electron is actually downward spinning. In a sense, quantum mechanics dictates that, until we actually measure the spin, we must consider the electron to possess only a tendency for an upward spin. If a subatomic reaction produces two electrons from a particle with no spin, one electron must possess an upward spin and the other a downward spin so that the net spin of the system remains zero. Yet the wave equation of each electron *must* contain terms corresponding to upward and downward spin. If, after the creation of the electron pair, these particles move far apart and then an observer measures the spin of one finding it to be "up," he can immediately conclude that the other electron, however far away, has a downward spin—in spite of the intrinsic ambiguity regarding its spin, which existed immediately before the measurement was performed on the first electron. It is almost as if the second electron acquired a downward spin at the time the spin of the first electron was measured. But how can observation have so profound an effect on the physics of the soldier in our analogy or on the physics of the second electron in the more realistic example? This is one of the great conundrums of modern physics.[2]

Several approaches to these difficulties have been proffered, but all eventually are based on the assumptions that perfect knowledge is impossible and that probabilistic, imperfect descriptions of nature are all that are available. Science must content itself with that pragmatic conclusion. This argument could be correct, but it is undeniably unsettling and has led many to the conclusion that quantum mechanics is as yet an incomplete discipline—one in need of both extension and overhaul.

There is another way out. It is a bizarre and startling way, but it is the only approach that permits a deterministic analysis of both the probability and the observer problems (if that kind of explanation is appealing). In its determinism it is a model of Western thought; in its conclusion, it is Eastern without doubt. In fact, we shall call it "Super Zen." [3]

This new approach is at heart a total return to mathematics, an absolute faith in the power of written logic. It asks for no sophisticated interpretation of the unknowable or of causality, no esoteric view of physics or metaphysics. It trusts the equations absolutely. If the equations say the soldier is both alive and dead, the soldier *is* both alive and dead.

But not in the same universe.

This is the startling conclusion arrived at by Hugh Everett III. All possible outcomes of all measured processes actually occur with the splitting of the universe (including the observer) instantaneously into separate cosmoses, in some of which the event occurred and in some of which it didn't. Travel or contact between these myriad universes is strictly forbidden by physics. We have no palpable knowledge of any universe but our own. Probability considerations determine in which universes the observer is likely to find himself and his colleagues. The expected outcome of an event plus "some form of" observation is statistical, not because of the basic nature of the processes of physics, but because the observer, like all events and objects, instantaneously splits and each of his copies is assigned to one or another universe—the number of which universes in any one state being determined by physical law. For example, if one were to place a highly radioactive material near a Geiger counter for one minute, the counter would, according to the laws of quantum mechanics, record at least one decomposition in, let us say, 10^{10} universes but not in one. We all would, with high probability, find ourselves in the universes in which the Geiger counter signals—and yet there is no inconsistency in a wave equation that suggests that in one universe it doesn't. (What life is like for those few copies of ourselves who are assigned to universes in which the improbable actually happens is left to the reader to imagine.) The wave equation encompasses all universes and provides the only way of glimpsing these other domains.[4]

According to this view, all possible genetic mutations *do* occur, all physically consistent life forms evolve, all possible outcomes of genetic engineering experiments occur, all chemically possible reactions take place. The statistical rule of the second law of thermodynamics need not, therefore, hold sway over all universes. Even as "our" universe is dominated by the law of increasing entropy, others remain unfettered by this restraint. Our "world(s)" are seen to be simply stones in the mosaic formed from the totality of all universes.

Using information theory and quantum mechanics, Everett has woven a theory relating not only these two disciplines but also the second law of thermodynamics and what might be called the theory of observation. Using Western deterministic methods, he has produced a world view—the so-called "many-worlds interpretation" of quantum mechanics—that is consistent with elements of Eastern thought. Super Zen not only provides a way to resolve basic questions in the interpretation of quantum mechanics, which parenthetically remains the basic tool for the description of the life process, but also provides an alternative view of the evolution, and indeed of the final state, of the universe(s).

Not many people take the "many-worlds interpretation" seriously—scientists allegedly have become more sophisticated about what kinds of knowledge they realistically expect from a physical theory like quantum mechanics, and the implications of "many-worlds" are so bizarre. But it is clear that the "many-worlds interpretation" proves science to be an open, questioning, free-wheeling, and very lively pursuit. More concretely, virtually everyone agrees that the more we know about the interpretation of quantum mechanics, the better off we all will be. The study of life is knocking on the door of quantum mechanics, and we will soon find that the fundamental precepts of this science have biological, and perhaps social or humanistic, impact—in one universe or another.[5]

CHAPTER XXI

Conclusion

Biology, like its subject matter, is alive. It is growing, maturing, changing, striving, hoping, and fearing. It is now undergoing what might be considered its adolescent growth spurt, bursting forth with exuberance and perhaps just a little self-consciousness and trepidation. What it needs is the maturation that hard work and responsible self-analysis can bring, as well as the benevolent direction of a good parent. Scientists will provide the work, both mental and physical, as well as their share of self-analysis and responsibility. Benevolent direction must come from society—and therein lies the great challenge.

Bioburst technologies potentially can increase crop yields, clean oil spills, and cure or prevent cancer, heart attack, stroke, and diabetes, as well as many other illnesses. Arthritis, the autoimmune diseases, and the inherited diseases of metabolism will eventually give way under its assault. The methods of molecular biology promise to clarify the processes involved in thought, growth, development, and even aging. The possibility that aging itself can be slowed or eliminated is real. Bioburst has already made it possible to conceive children in one year and bear them in another. Electronic microsurgery could provide medicine with safe access to any part of the human anatomy. Scarce hormones have, and will, be made abundant. For the first time, direct therapy of viral infections appears to be possible and in some cases has already been accomplished. New diseases can be expected to yield to the therapeutic assault of science soon after they arise.

Even more important, perhaps, is the insight that man will gain into the nature of life and into his own nature as well. This, coupled with an ability to affect the direction of life forces, will change our lives. Although some may argue that these developments desanctify life, it appears more appropriate to conclude that they expand and enhance man's appreciation of it. There is no doubt that man's view of life will change. In fact, no greater alteration in the human perception of life and its uniqueness appears likely to occur in the future, except perhaps the intellectual adjustment that would be required were life to be found on other worlds. But, like that discovery, the insights provided by Bioburst enhance and glorify life rather than cheapen it. The old view of life may no longer remain sacred, but life itself will become more cherished.

Like all human activities, molecular biology is associated with risk. Of course, not taking advantage of molecular biology is also a risk, as is evident from the fact that, theoretically, the appearance of a new infectious disease could at any moment destroy the entire human race. The only defense we have against such hypothetical plagues is biological science. Although a man-made plague could potentially result from Bioburst technology, this possibility seems extremely remote if appropriate precautions are taken.

John Naisbitt has noted in *Megatrends*—his analysis of social trends in society—that new technologies are often introduced in patently benign and unobtrusive forms. Robots first appeared in America as toys for children; only later were they found making automobiles in a Detroit assembly line. The risks and potential benefits of a new technology cannot be exactly predicted when that technology first makes its appearance. Thus, continued oversight and analysis over time appear to be the appropriate societal responses to a powerful new technology. Nonetheless, it appears that if the cloning vectors now commonly used—or other vectors with similar safety characteristics—continue to be the mainstay of Bioburst activities, society is reasonably safe. To the extent that pathogens are manipulated, strict oversight must be exercised, and such activity should be kept to a minimum. In the case of the genetic therapy of human disease, the potential risk, on a societal as opposed to an individual level, seems small in the case of somatic cell therapy and manageable in the case of germ line therapy. In the latter instance, the dose of engineered genes in the human species may well be a critical variable and should be monitored in an ongoing fashion if this kind of therapy is ever undertaken.

The free release of genetically engineered organisms is similarly associated with risk and requires even more oversight, planning, and monitoring than the elementary applications of Bioburst technology to the treatment of human disease, since any associated ecological changes must be monitored along with the dose of engineered organisms in any given species. This kind of oversight can only be provided if the right of the entrepreneur to apply his newly invented organism (albeit for the good of society) is understood to be explicitly outweighed by society's need for ecological security. This balancing of interests will provide a great social challenge, but the process must not, and should not, assume an adversarial character. Indeed, removing the intensely adversarial nature of so many government-industry and government-individual interactions in this nation while maintaining a healthy respect for the rights of all interests is emerging as a great challenge for social planners. It is a particularly important issue in the case of free-release genetic engineering projects. Finally, because life per se does not respect international boundaries, any regulation or monitoring of bioengineering should ideally be standardized on an international level. Clearly, the pressure to use, for example, the free release of engineered organisms to enhance crop yield will be more severe in famine-ridden Ethiopia than in abundant America, yet the world as a whole must be protected from any reckless use of new life forms. Enlightened oversight should encompass the entire world.[1]

Can science, industry, government, and the public at large be trusted to put aside corrupting avarice in order to seek the high ground of enlightened self-interest? Jeremy Rifkin concludes, like others before him, that Darwinian evolution was more a product of the social context in which it developed than of legitimate science. From this he extrapolates that all science and all philosophical premises directly or indirectly derived from it are of necessity the products of social context, and as such they are not to be trusted. The implication is that Bioburst must stop short of any major application (perhaps even stop entirely) lest it be used in a manner that appears appropriate in the context of one society or situation but is inappropriate when observed over all time and space.[2]

Yet science cannot be totally corrupt. Although some may contend that the great scientific advances were as much the result of sales-manship as of experimentation, and although scientists may be en-countered who fabricate data, nonetheless airplanes fly, electric motors whirl, nuclear power plants generate electricity, plant graft-ing produces new crops, vaccination prevents disease, rockets orbit the earth, and antibiotics cure disease. Science is alive, vital, and essentially honest. It is also cultural, but that should not be con-strued as an indictment. As Reinhold Niebuhr has written,

> The individual is the product of the whole socio-historical process, though he may reach a height of uniqueness which seems to transcend his social history completely. His individual decisions and achievements grow into, as well as out of, the community and find their final meaning in the community. Even the highest forms of art avail themselves of tools and forms, of characteristic insights and styles which betray the time and place of the artist; and if they rise to very great heights of individual insight, they will also achieve a corresponding height of universal valid-ity. They will illustrate, or penetrate into, some universal, rather than some particular and dated experience, and thereby will illuminate a life of a more timeless and wider community.[3]

Culture nurtures science and science nurtures culture. The fact that science exists in a cultural context is not a priori an indictment, but the cultural misuse and misinterpretation of scientific data and hy-potheses, particularly as applied to social issues, can pose risks, both psychic and physical.

Few opinion shapers, political leaders, industrialists, or citizens at large really understand the fundamental nature of science, which is not and cannot be arrogant or arbitrary—although some of its practitioners are. Science deals with observations and models, not with absolutes. It comes as a shock to many people that two of the most heralded advances in twentieth-century science, quantum me-chanics and relativity theory, have never been adequately recon-ciled. Einstein himself was skeptical of the underpinnings of quan-tum mechanics, yet that science has been extremely productive both in its insights and its practical applications. Science is an ever-evolving array of intellectual models, not a rigid structure. Once this realization (be it derived from a particular social context or not) is accepted, there need no longer be a philosophical fear of science, only vigilance, caution, rational contemplation, and openness to new ideas. Clear thinking based on constant questioning of premises is what is required. We must all adopt the mode of thinking that is so well typified by the modern-day physicists who wonder if the gravitational "constant" G is changing over time. Although many of us may never before have felt the need to think in this way, Bioburst, abutting as it does on human life itself, compels us to adopt a more sophisticated mode of thought. For his own mental health, not to mention his greater appreciation of the world around him, man must come to terms with the basic nature of science. He must be comfortable with the known, the unknown, and the unknowable. He must blend scientific theories with personal spiritual beliefs. Modern man is compelled to become a practicing philosopher.[4]

Even this is not enough if Bioburst is to be harnessed as a benevolent force. The average man must learn more about the nature of biology, for he is a part of biology. Although great technical expertise is not required, a basic understanding of the processes and risks involved in Bioburst applications is essential. This need for technical knowledge has also been present at the time of the introduction of other technologies, but in the past society has left decisions to industrial or state experts who often were not motivated to investigate potential problems. Now we live in an era of ever-increasing participatory democracy and decisions are made more collectively. Such decisions require information (read "education"). If no one outside the nuclear community had heard of "meltdown," the anti-nuclear movement would not be as strong as it is today. Indeed, the nuclear power issue is a prototype for the discussion of Bioburst technology. Some years ago, national energy experts wrote, "The public will have to choose between energy sources based on individual values and beliefs about social ethics—not on the advice from technical experts."[5] This will be true in the case of Bioburst as well. To accomplish this task, the public must understand the nature of science, the nature of biology, and the goals of the new technology.

The naïve faith that bioengineers will make all life "better," and the resulting counterargument that no one knows what "better" is, will have to be seen for the shams that they are. In biology, there is no better or worse. There is only life, which must be cherished but which always changes.

The great challenge of Bioburst, and its great hope, like the challenge and hope of past scientific accomplishments, lie at the feet of the individual. If enlightened oversight is to be achieved, we must undertake a tremendous education of society, and we must all become more contemplative and vigilant. There is comfort, however, in the realization that there were those in the past who proved it possible to be scientifically and philosophically aware while being quintessentially human. We must listen to their voices:

> A crazy old man that is near 70 having lately enjoy'd the benefits of inoculation 'tis thought that if he should happen to die one minute before 90, these people (if not come to their senses before) will say this inoculation killed him.
>
> But to what purpose is all this jargon? And of what significancy are most of our speculations? Experience! Experience! 'Tis to thee that the matter must be referr'd after all. A few empericks here, are worth all our dogmatics.
>
> (Cotton Mather, 1722)[6]

> Is man, then, a defenseless and perplexed victim at the mercy of the irresistible superior power of technology? He would be if man today abandons any intention to pit meditative thinking decisively against merely calculative thinking. But once meditative thinking awakens, it must be at work unceasingly and on every last occasion.
>
> (Martin Heidegger, 1955)[7]

> Shall not physicians and surgeons recommend and bring it into greater esteem and practice, and save (under God) thousands and tens of thousands by it; and make further improvements in it.
>
> (Zabdiel Boylston, 1730)[8]

Enzymes and the Laws of Thermodynamic Action

To appreciate life, one must appreciate enzymes, which make possible the chemistry of life. And in order to understand the functioning of enzymes, it is necessary to appreciate that all chemical reactions proceed in accordance with the laws of thermodynamics and, in particular, with the second law. Briefly stated, the first law of thermodynamics concerns the conservation of energy. If q is taken to indicate the heat (*i.e.*, energy) added to a system and u the total energy of the system, with w denoting work done by the system (*i.e.*, energy expended), the first law of thermodynamics states that $q = \Delta u + w$, where Δu indicates the change in total energy u. Of the work done by the system, a portion, w', can be easily recovered for constructive purposes, and a part, w'', is often associated with changes in the volume of the system. Therefore, a new, albeit artificial, term can be coined—*enthalpy,* or "heat content"—so that the change in enthalpy, Δh, is equal to the change in total energy plus the volume work: $\Delta h = \Delta u + w''$. Substituting into our first equation, we have $q = \Delta h + w'$, where w' indicates readily recoverable work.[1]

The first law of thermodynamics is not a complete description of reality, since it in no way determines the direction of a reaction. For example, if one were to drop an egg on the floor, the egg would gain energy in the fall and would then distribute this energy to its pieces when it smashed. The reaction would read: egg on table + kinetic energy → pieces of egg scattered on floor. The egg while on the tabletop would have associated with it potential energy $e = gmh$, where m is the mass of the egg, h is the height of the table, and g is the acceleration of gravity. Once the egg is pushed over the edge, this potential energy is converted to kinetic energy, which eventually breaks the egg ("does work on the egg") on the floor. The pieces of egg would be at lower potential energy following the fall and work would have been "done." There is nothing in the first law that precludes the idea that the same energy (gmh) delivered to the pieces of egg on the floor could accomplish the reverse reaction (pieces of egg on the floor + energy $gmh \rightarrow$ egg on table), yet this can never occur spontaneously. A broken egg cannot be put back together by adding gmh as heat energy. This is so because all chemical processes (at least at this stage of the development of the universe) tend toward equilibrium or "randomness."

This tendency toward randomness is enunciated in the second law of thermodynamics. Classically, this law is expressed in terms of *entropy* (S), which is defined so as to give a measure of the degree of randomness or disorder in a system. Mathematically, $S = S_0 + {}_a^b\int dq/T$, where S_0 is an arbitrary starting constant equal to the entropy of the system in state a and dq/T is the change in the ratio of heat to temperature for the transition of the system from state a to state b—if carried out exclusively via states of equilibrium (*i.e.*, states of maximum randomness). In other words, if any given system were to be moved from one state to a second state employing maximally equilibrated conditions, the order forever lost to randomness in the process is given by S. In reactions performed at a constant temperature, the product of $S \times T$ is then a measure of the energy that cannot be recovered and is irretrievably lost to the continued increase in randomness in the universe. That is, this energy is totally lost for useful purposes, imparted to the random motion of molecules and thus becoming part of the general movement toward equilibrium or randomness going on all about us. No matter how the system is taken from state a to state b, a certain amount of randomness will inevitably be associated with the process, and the second law will permit us to determine how much energy thereby is forever lost from the potential of doing useful work. Formally, the second law of thermodynamics states that any closed system will spontaneously evolve in the direction of increasing entropy. Once a system reaches equilibrium, S is at a relative maximum. If a system in equilibrium is heated, the added energy will be imparted to the system in the form of increased random motion (*i.e.*, it will be unrecoverable). Therefore, if a system at equilibrium is heated, $q = \Delta H + w'$ but $q = \Delta ST$. Thus, $\Delta H = \Delta ST - w'$. If a system is not at equilibrium, q does not equal ΔST but is something less than ΔST because, in a system not at equilibrium (*i.e.*, not at maximum randomness), S can increase as the system approaches equilibrium and in fact will tend to increase spontaneously even without added heat. Therefore, if a system is not in equilibrium, $\Delta H < \Delta ST - w'$ and $\Delta H - \Delta ST < -w'$. If we now define free energy (G) as $G = H - ST$, we can come to some interesting conclusions. If temperature is constant, as it usually is in mammals and biochemical test systems, $\Delta G = \Delta H - \Delta S(T) < -w'$. As the system approaches equilibrium, $\Delta H - \Delta S(T)$ and therefore ΔG approach $-w'$. Thus, the amount of useful energy that can be derived from a system as it approaches equilibrium is $-\Delta G$, since in this case $\Delta G = -w'$. Furthermore, reactions spontaneously move in directions so that free energy decreases.

For any chemical reaction in which $A + B \rightleftharpoons C + D$, it can be shown that $\Delta G = -RTlnK$, where R is the gas constant; K, the equilibrium constant, equals $[C][D]$ divided by $[A][B]$ at equilibrium; and $[C][D][A]$ and $[B]$ are the molar concentration of these reactants.

The second law of thermodynamics then tells us that if for any given reaction the ΔG is negative, the reaction can proceed spontaneously. On the other hand, if ΔG is positive, the reaction cannot occur unless energy is provided to drive the reaction. Additionally, the second law tells us that the total entropy of the universe is increasing. Any local decrease in entropy produced by the infusion of energy must be offset by increases elsewhere in the universe.

The statement that "if ΔG is negative the reaction can proceed spontaneously" does not mean that the reaction *will* proceed spontaneously with any rapidity. As noted earlier, the combustion of glucose to CO_2 and H_2O is associated with negative free energy and can therefore proceed spontaneously and at the same time provide useful energy. Indeed, this overall reaction is the source that supplies the human body with a major portion of its energy needs. Yet glucose does not spontaneously undergo combustion on the kitchen table. A match must be placed to a spoonful of sugar in order to start the reaction and release the free energy of the system. The same effect can be observed in other systems. There is enormous free energy in the fission of uranium[235] into daughter elements, and the reaction does proceed spontaneously to some extent. It proceeds a great deal faster if a small amount of energy is added in the form of neutron collisions, which then lead to a chain reaction.

What are the energistics, if one will, of the burning of carbohydrate? They may be graphically depicted as shown in figure 26. Activation energy must be supplied to begin the reaction (compare, for example, priming a pump). Reaction energy could be provided by increased temperature, but this would have dire consequences for the body if it were actually used to any extent by biological systems. However, some enzyme systems of certain "cold-blooded" species are functionally temperature-dependent under normal conditions, leading, therefore, to such temperature-dependent phenomena as the nocturnal lethargy of snakes. As an alternative to supplying reaction energy as heat, enzyme systems could be used to reduce activation energy to a level achievable at body temperature. Thus, our energy plot in the presence of an enzyme system could look as shown in figure 27 for the metabolism of glucose.

Additionally enzyme systems permit the released energy to be captured as ATP, rather than allowing the combustion of carbohydrate to be wasted as heat. Conversely, when energy must be provided to a system in order to drive a reaction, enzyme systems see to it that energy is specifically delivered where needed and not randomly applied to the whole cell. If energy is available from ATP to reverse adverse energistics, enzymes can do a great deal to focus this energy where it is needed. Thus, enzymes can be seen to be very powerful modifiers of biochemical reactions. But they are not omnipotent. Energy must be applied in sufficient amount to reverse the unfavorable energistics of any reaction. If a reaction is so unfavorable that adequate energy is unavailable to reverse it, no enzyme can make the reaction begin.

If entropy spontaneously must increase, how could life have evolved? We could also ask what set the stage for the formation of the stars and galaxies and the nuclei of heavy metals in a cosmos characterized by increasing entropy. The classical answer to the first

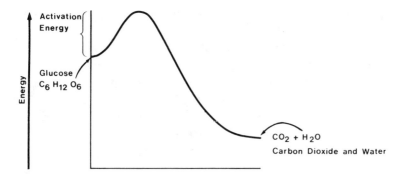

Fig. 26. Energistics of the burning of carbohydrate.

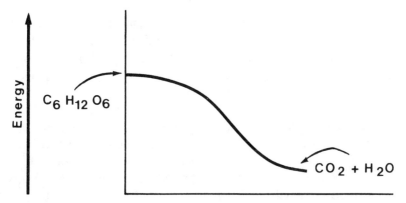

Fig. 27. Metabolism of glucose in the presence of an enzyme system.

question seems relatively straightforward. The free energy for the formation of life from carbon, hydrogen, oxygen, nitrogen, and other elements appears to be positive, and therefore this process could only have occurred in the past, and indeed can only continue to occur now, with the provision of external energy. Fortunately there was, and is, such a source of energy—the sun. All life is dependent on the energy of the sun. Heat and light from the sun are captured by the process of photosynthesis, leading to the production of sugars by plants, which are then metabolized to ATP, which in turn provides the power to make complementary DNA (life) out of nucleotides. Prior to the development of the photosynthesis process by biological systems, heat energy, together with energy-producing chemical reactions catalyzed by clays or other inorganic materials, probably served the same role.[2]

Were the sun to stop shining, either an alternative form of energy would have to be found to support life or life would end. In lay terms, we would freeze and starve—but actually it would be the first and second laws of thermodynamics that did us in. The energy of the sun permits the entropy of living things to decrease while the entropy of the sun increases. It must be recalled that the total entropy of the system *sun + life on earth* must increase, even if the entropy of living things transiently decreases. The second law cannot be thwarted.

The second question, and indeed a more complete response to the first, is not very straightforward, and no definitive answer can be given. As long as we consider the earth (including life) and the sun as a system, we can hypothesize that the decrease in entropy associated with the advent and perpetuation of life from nonliving chemicals is associated with an even greater increase in entropy of the sun. The sun can be considered to be "running down" and to be supporting life on earth in the process. This analysis, of course, begs the question of how *likely* is the emergence of life in a world of increasing entropy. According to thermodynamic principles, it appears to be extremely unlikely. It is probably safe to say that the genesis of life has not yet been fully reconciled with the second law. And what about the cosmos as a whole? How did stars and galaxies form from nuclear plasma? Why is intrastellar gas still forming stars? To put this another way, how did the universe begin with or acquire a low entropy, which it then had to increase while permitting the appearance of small islands of dramatically decreasing entropy, such as the niche occupied by life?[3]

The answer is anything but obvious, and the last word has certainly not yet been said on this matter. One theory holds that the tremendous energy of the presumed Big Bang that gave birth to the universe resulted in an enormous expansion of time and space. In this process, the grand unifying force predicted by modern physics gradually "froze out" as temperatures decreased, revealing the multiple facets of itself that we now know to be the basic forces of nature, such as electromagnetism and the strong and weak nuclear forces. But even before the Big Bang, just after the embryonic microcosmos "quantum tunneled" into existence from nothing, alterations were produced in the parameters responsible for defining the nature of gravitational interactions. This had the effect of producing an extremely rapid inflation, or exponential expansion, which, in turn, was followed by the Big Bang and a slower, sustained expansion of the universe. The inflationary period followed by the expanding universe led to a relative (when compared to the maximum possible entropy) *decrease* in entropy as well as to a disparity between the entropy associated with gravitational fields as compared to that associated with matter and radiation. As the universe cooled, these eras of decreasing entropy came to an end, leading to the present epoch during which net entropy is still increasing in order to offset, or pay back, the decreased entropy of an earlier era—this paying back, for example, could account for the continued stellar synthesis of heavy elements from hydrogen, the continued (at least partially gravity-mediated) clumping of interstellar material to form stars and galaxies, and other processes that increase entropy as the universe tends, however slowly, to "run down." The net result is the current overall tendency toward increasing entropy, which provides, as Sir Arthur Eddington pointed out, the "arrow of time." We perceive time as flowing in the direction of increasing entropy, and the fact that our universe at the present moment is in an era of increasing entropy imparts a sense of time to our world.[4]

It must be pointed out that not all authors agree with this explanation of the origin of increasing entropy and the present direction of time, so it is probably safest to say that a definitive explanation for the formation of galaxies, much less of life, in a cosmos in which time's arrow appears to be pointing in the direction of increasing entropy is currently unavailable. This is one of the great cosmological questions. However, to further confuse the issue, it also must be noted that the second law of thermodynamics (in either its classical or its statistical mechanical form), like any physical law, need not necessarily apply at all eras in the past and future and at all points in space. After all, human science is only a few thousand years old, so the constancy of our physical laws over periods of billions of years, although supported by considerable indirect evidence, must nonetheless be taken as an act of faith. Again we have touched upon a field that is open for theoretical discussion.[5]

Notes

In citing journals in the notes, short titles have generally been used. Journals frequently cited have been identified by the following abbreviations:

AIM   *Archives of Internal Medicine*
AJC   *American Journal of Cardiology*
AN   *Acta Neuropathologica* (Berlin)
JAMA   *Journal of the American Medical Association*
JBC   *Journal of Biological Chemistry*
JCI   *Journal of Clinical Investigation*
NEJM *New England Journal of Medicine*
PNAS *Proceedings of the National Academy of Sciences*
RMP   *Reviews of Modern Physics*

Chapter I

  1. Otho T. Beall, Jr., and Richard A. Shryock, *Cotton Mather: First Significant Figure in American Medicine* (Baltimore, 1954); Barrett Wendell, *Cotton Mather: The Puritan Priest* (Cambridge, Mass., 1926), 17.

  2. Wendell, *Cotton Mather*, 35–37.

  3. *Ibid.*, 99, 101, 102.

  4. *Ibid.*, 106.

  5. *Ibid.*, 8; Increase Mather, *Several Reasons Proving that Innoculating or Transplanting the Small-Pox, Is a Lawful Practice, and that It Has Been Blessed by God for the Saving of Many a Life,* and Cotton Mather, *Sentiments on the Small-Pox Innoculated,* reprinted from the original folio single sheet printed in Boston, 1721, with an introduction by George Lyman Kittredge (Cleveland, 1921), 5.

  6. Lady M. W. Montagu, *The Letters and Works of Lady Mary Wortley Mon-*

tagu, ed. James Archibald Wharncliffe (2 vols.; New York, 1893), I, 308; R. W. Clark, *Benjamin Franklin* (New York, 1983), 18; Cotton Mather, *An Account of the Method and Success of Inoculating the Small-pox in Boston in New-England* (London, 1722), 7.

7. Zabdiel Boylston, *An Historical Account of the Smallpox Inoculated in New England, upon All Sorts of Persons, Whites, Blacks, and of All Ages and Constitutions* (2nd ed.; London, 1730), 39–45; C. Mather, *An Account of the Method and Success*, 19. Mather notes that Boylston chose to leave the patients "to the Liberties" after inoculation (p. 9).

8. C. Mather, *An Account of the Method and Success*, 10. Mather commented that the attempt of the selectmen of Boston to indict Boylston for his inoculation procedure was intrinsically hypocritical, since it was considered perfectly legal and moral for parents to introduce young children into the sickrooms of patients suffering with smallpox so that they might be infected at an early age when the disease presumably would be mild. The selectmen's basic charges against Boylston were that his procedure could cause death, that it involved the inoculation of filth into blood and could thus be considered corrupting, and that inoculation tended to spread and continue the infection in a place longer than it might otherwise persist. Mather attempted to refute these arguments (*ibid.*, 23–24).

9. Wendell, *Cotton Mather*, 279.

10. Boylston, *An Historical Account*, 40.

11. C. Mather, *An Account of the Method and Success*, 22, 23, 24. For additional justifications provided by the Mathers for the practice of inoculation, as well as to uncover a treasure trove of information regarding the occurrences of 1721, see I. Mather, *Several Reasons Proving*, and C. Mather, *Sentiments on the Small-Pox*.

12. Cotton Mather, *A Reply to the Objections Made Against It from Principles of Conscience*, in Benjamin Colman, *A Narrative on the Method and Success of Inoculating the Small-Pox in New-England* (Dublin, 1722).

13. C. Mather, *An Account of the Method and Success*, 7.

14. Beall, *Cotton Mather*, 126; C. Mather, *An Account of the Method and Success*, 44.

Chapter II

1. James D. Watson, *The Molecular Biology of the Gene* (3rd ed.; Menlo Park, Ca., 1976), 1–83.

2. T. O. Diener, "Viroids and Their Interactions with Host Cells," *Annual Review of Microbiology*, 36 (1982), 239–58; T. O. Diener, "Viroids as Disease Agents," *National Cancer Institute Monograph*, 60 (1982), 161–67; T. O. Diener *et al.*, "Viroids and Prions," *PNAS*, 79 (1982), 5220–24; T. Kiss, "Sequence Homologies Between a Viroid and a Small Nuclear RNA (snRNA) Species of Mammalian Origin," *Federation of European Biochemical Societies Letters*, 144 (1982), 318–20; J. Haseloff *et al.*, "Viroid RNAs of Cadang-Cadang Disease of Coconuts," *Nature*, 299 (1982), 316–21; E. Dickson, "A Model for the Involvement

of Viroids in RNA Splicing," *Virology,* 115 (1981), 216–21; T. O. Diener, "Viroids," *Scientific American,* 244 (1981), 66–73; A. D. Branch and H. D. Robertson, "A Replication Cycle for Viroids and Other Small Infectious RNAs," *Science,* 223 (1984), 450–55; P. Van Wezenbeek *et al.,* "Molecular Cloning Characterization of a Complete DNA Copy of Potato Spindle Tuber Viroid RNA," *Nucleic Acids Research,* 10 (1982), 7947–57; D. C. Gajdusek, "Unconventional Viruses and the Origin and Disappearance of Kuru Signs," *Science,* 197 (1977), 943–60; H. Diringer, "Scrapie Infectivity Fibrils and Low Molecular Weight Proteins," *Nature,* 306 (1983), 476–78; R. H. Kimberlin, "Scrapie Agent: Prions or Virinos," *Nature,* 279 (1982), 107–108; R. G. Rohwer, "Virus-like Sensitivity of the Scrapie Agent to Heat Inactivation," *Science,* 223 (1984), 600–601; P. A. Merz *et al.,* "Scrapie-Associated Fibrils in Creutzfeldt-Jakob Disease," *Nature,* 306 (1983), 474–76; P. Brown *et al.,* "Chemical Disinfection of Creutzfeldt-Jakob Disease Virus," *NEJM,* 306 (1982), 1279–82; P. Duffy *et al.,* "Possible Person to Person Transmission of Creutzfeldt-Jakob Disease," *NEJM,* 290 (1974), 692–93; S. B. Prusiner, "Novel Proteinacious Infectious Particles Cause Scrapie," *Science,* 216 (1982), 136–44; E. McKinley *et al.,* "A Protease-Resistant Protein Is a Structural Component of the Scrapie Prion," *Cell,* 35 (1982), 57–62; R. G. Rohwer, "Scrapie Infectious Agent Is Virus-like in Size and Susceptibility to Inactivation," *Nature,* 308 (1984), 658–62. Perhaps scrapie is caused by a small virus after all. See C. L. Master *et al.,* "The Familial Occurrence of Creutzfeldt-Jakob Disease and Alzheimer's Disease," *Brain,* 104 (1981), 535–58; Jaap Goudsmit *et al.,* "Evidence For and Against the Transmissibility of Alzheimer's Disease," *Neurology,* 30 (1980), 945–50; S. B. Prusiner, "Some Speculations About Prions, Amyloid and Alzheimer's Disease," *NEJM,* 310 (1984), 661–63.

3. Lewis Thomas, *The Lives of a Cell: Notes of a Biology Watcher* (New York, 1974), 81. In this essay, "Organelles as Organisms", Thomas discusses the suggestion that chloroplasts and mitochondria are endosymbionts and further muses on the impact of this concept on the psyche of man.

4. See, for example, E. P. Wigner, *Symmetries and Reflections: Scientific Essays* (Cambridge, Mass., 1970). For more on this subject, see pp. 179–80.

5. A. White, P. Handler, and E. Smith, *Principles of Biochemistry* (3rd ed.; New York, 1964)—a readable discussion of basic biochemistry. The four bases of DNA are adenine, thymine, guanine, and cytosine. When linked to deoxyribose, these bases form the four nucleosides deoxyadenosine, deoxythymidine, deoxyguanosine, and deoxycytidine, which in the text are referred to as adenosine, thymidine, guanosine, and cytidine. The corresponding nucleotides are deoxyadenosine-5'-phosphate, deoxythymidine-5'-phosphate, deoxyguanosine-5'-phosphate, and deoxycytidine-5'-phosphate.

6. G. Kolata, "Drug Transforms Transplant Medicine," *Science,* 221 (1983), 40–42.

7. For a succinct discussion of the evolution of the one gene–one enzyme concept, including a discussion of the roles of Mendel, Garrod, Johannsen, Beadle and

Tatum, and Jacob and Monod, see J. B. Stanbury, J. B. Wyngaarden, and D. S. Fredrickson (eds.), *The Metabolic Basis of Inherited Disease* (3rd ed.; New York, 1972), 3.

8. N. Proudfoot, "The End of the Message and Beyond," *Nature*, 307 (1984), 411–13 (discusses advances in our understanding of the transcription-translation process and describes recently discovered bits of RNA that have the capacity to function as enzymes during intron processing and message termination); A. Nordheim and A. Rich, "Negatively Supercoiled Symian Virus 40 DNA Contains Z-DNA Segments Within Transcriptional Enhancer Sequences," *Nature*, 303 (1983), 674–79 (investigates the possibilities that structural changes in DNA could represent control mechanisms for RNA transcription).

Chapter III

1. J. Keosian, *The Origin of Life* (New York, 1964). "Meteoric Rise," an article in the New York *Times,* January 8, 1984, Sec. E, p. 6, describes the recent finding that the Murchison meteor that fell in Australia in 1969 contains the five chemical bases required for the human genetic code and mentions that scientists have been able to synthesize all five from a primordial chemical soup. These two observations suggest that both in outer space and on the early earth the building blocks of the human genetic code could have formed spontaneously. Lawrence J. Henderson, in *The Fitness of the Environment: An Inquiry into the Biological Significance of the Properties of Matter* (Boston, 1958), 297, suggests that "if it is life's essential aim to catch up unstable energy in order to expend it in explosive actions, it probably chooses, and on each solar system and on each planet, as it does on earth, the fittest means to get this result in the circumstances with which it is confronted." The recent description of Archaebacteria—organisms found on the ocean's floor that are capable of living in extremely warm temperatures and of subsisting on what otherwise might be considered noxious compounds—have expanded the traditional definition of a "fit" environment. See F. Fisher *et al.,* "Chemolithoautrophic Metabolism of Anaerobic Extremely Thermophilic Archaebacteria," *Nature*, 301 (1983), 511–13. A. G. Cairns-Smith, "The First Organisms," *Scientific American,* 252 (1985), 90–100, argues that the earliest life forms may have been self-reproducing bits of crystalline clay that over time developed, perhaps for purposes related to improved crystal structure, the capacity to catalyze the synthesis of nucleotides. Thereafter the nucleotides developed their own biology. Although the five bases of the genetic code have been found in the Murchison meteor and therefore presumably could have been present on the early earth, the way in which nucleotides came to exist in abundance remains unknown. Synthesis catalyzed by clays is an interesting possibility.

2. T. W. Driker, J. J. Bujarski, and T. C. Hall, "Mutant Viral RNAs Synthesized *in Vitro* Show Altered Amnioacylation and Replicase Template Activities," *Nature,* 311 (1984), 171–75. In actual fact, several plant viruses contain RNA genomes, the 3' ends of which exhibit some enzymelike functions. Recently it has been

shown that RNA can in certain circumstances serve as a catalyst, much as enzymes do. See R. Levin, "RNA Can Be a Catalyst," *Science,* 218 (1982), 872–74; H. Kuhn and J. Wasser, "Evolution of Early Mechanisms of Translation of Genetic Information into Polypeptides," *Nature,* 298 (1982), 585–86 (an intriguing discussion of the RNA hairpin loop and the early development of translation by life forms).

3. J. Cherfas, *Man-Made Life: An Overview of the Science, Technology and Commerce of Genetic Engineering* (New York, 1983), 226. In actual fact, not all divergent genetic codes disappeared. Mitochondria still seem to preserve a few coding variations, but all works out well because *their* tRNAs understand *their* codes.

4. R. N. Re, "The Cellular Biology of the Renin-Angiotensin Systems," *AIM,* 144 (1984), 2037–2041; W. E. Stumpf *et al.,* "Target Cells for 1,25-dihydroxy-vitamin D_3 in Intestinal Tract, Stomach, Kidney, Skin, Pituitary, and Parathyroid," *Science,* 206 (1979), 1188–90. These two papers describe the existence of renin in the first case, and vitamin D sensitive cells in the second case, in body tissues not heretofore appreciated by medicine as being involved with these substances. In fact, the evolutionary development of the human being from a colony of smaller cells almost of necessity implies that chemical compounds used systemically in the intact animal would have local functions at the cellular level far different from those that macroscopic medicine recognizes. Not only these studies but many others confirm this point and also lead to interesting new approaches to therapy. For example, vitamin D analogues can in certain blood cells turn off oncogene transcription resulting in the differentiation of undifferentiated cancer cells. How life came to exist is unknown. Among theories currently under study are: (1) the possibility that nucleic acid and protein polymers formed in the primordial seas; (2) the idea that the first life may have been protein-based, leading to the subsequent development of nucleic acid–based heredity; (3) the hypothesis that the first self-replicating "organisms" were clays that subsequently catalyzed the chemical reactions needed for the genesis of life as we know it; and (4) the possibility that life first developed elsewhere in the cosmos and then traveled through space to seed the earth. Any theory of the origin of life must eventually answer two critical questions. First, how can the genesis and evolution of life be reconciled with the second law of thermodynamics? And, what were the specific chemical steps by which life arose? See Richard J. Field, "Chemical Organization in Time and Space," *American Scientist,* 73 (1985), 142–50; and D. A. J. Tyrrell, "A New Dimension to Evolutionary Theory?" *Nature,* 294 (1981), 489–90.

Chapter IV

1. P. Leder, "The Genetics of Antibody Diversity," *Scientific American,* 346 (1982), 102–106.

2. Stephen J. Gould, *The Panda's Thumb: More Reflections in Natural History* (New York, 1980), 76. For an interesting view of the tragic interaction between

217

failed theory and failed politics, see Stephen J. Gould, *Hens' Teeth and Horses' Toes: Further Reflections in Natural History* (New York, 1983), 134–44.

3. S. Tonegawa, "Somatic Generation of Antibody Diversity," *Nature*, 302 (1983), 575–81; J. G. Seidman, E. E. Max, and P. Leder, "A K-Immunoglobulin Gene Is Formed by Site-Specific Recombination Without Further Somatic Mutation," *Nature*, 280 (1979), 370–75; P. Leder, "Genetic Control of Immunoglobulin Production," *Hospital Practice* (February, 1983), 73–82; P. A. Hieter *et al.*, "Clustered Arrangement of Immunoglobulin Lamba Constant Region Genes in Man," *Nature*, 298 (1981), 536–40.

4. "Nobel Prize to Barbara McClintock," *Nature*, 305 (1983), 575.

5. P. Newmark, "Prizes (At Last) for Immunology," *Nature*, 311 (1984), 601, discusses the work of Georges Kohler and César Milstein as well as that of Niels Jerne, all of whom shared a Nobel Prize in 1984. See the classic paper by Kohler and Milstein, "Continuous Cultures of Fused Cells Secreting Antibody of Predefined Specificity," *Nature*, 256 (1975), 495–97.

6. In order to appreciate how difficult the production of useful, specific antiserum is in the absence of monoclonal technology, see, for example, J. Nussberger *et al.*, "Selectivity of Angiotensin II Antiserum," *Journal of Immunologic Methods*, 56 (1983), 85–96. This work, in which I participated, involved the synthesis of a large number of modified angiotensin II chemical analogues and the immunizing of an even larger number of animals in order to develop a few sera with interesting properties for use in radioimmunoassay.

7. R. A. Miller *et al.*, "Treatment of B-Cell Lymphoma with Monoclonal Anti-Idiotype Antibody," *NEJM*, 306 (1982), 517–22; R. A. Reisfeld, "Monoclonal Antibodies to Human Malignant Melanoma," *Nature*, 298 (1982), 325–26; S. M. Larson *et al.*, "Diagnostic Imaging of Malignant Melanoma with Radiolabeled Anti-Tumor Antibodies," *JAMA*, 249 (1983), 811–12; J. Q. Trojanowski and V. M.-Y. Lee, "Anti-Neurofilament Monoclonal Antibodies: Reagents for the Evaluation of Human Neoplasms," *AN*, 59 (1983), 155–58; S. M. Larson, "Localization of ^{131}I-Labeled $_p$97-Specific Fab Fragments in Human Melanoma as a Basis for Radiotherapy," *JCI*, 72 (1983), 2101–2114 (this landmark study describes a large-scale deliberate attempt to use monoclonal antibodies labeled with radioactive tags for the treatment of human neoplastic disease); I. R. Mackay *et al.*, "Germ-line Deletion of Genes Coding for Self-Determinants," *Nature*, 288 (1980), 302–303 (discusses the possibility that germ line gene deletions could be responsible for the body's recognition of "self" from "foreign"—the hypothesis in general seems weak).

Chapter V

1. P. Rous, "Transmission of Malignant New Growth by Means of a Cell-Free Filtrate" (1911), reprinted in *JAMA*, 250 (1983), 1445–46, with commentary by Henry C. Pitot, pp. 1447–49; R. C. Gallo, "The Virus-Cancer Story," *Hospital Practice* (June, 1983), 79–89. Z. Trainin *et al.*, "Suppression of the Humoral

Antibody Response in Natural Retrovirus Infections," *Science,* 220 (1983), 858–59; R. C. Gallo and F. Wong-Staal, "Retroviruses as Etiological Agents of Some Animal and Human Leukemias and Lymphomas and as Tools for Elucidating the Molecular Mechanisms of Leukemogenesis," *Blood,* 60 (1982), 545–57. An interesting curiosity is that hepatitis B virus, which in certain circumstances can predispose to hepatitic cancer, is a DNA virus that is associated with a reverse transcriptase activity—which in turn suggests that this potentially oncogenic DNA virus utilizes an RNA intermediate in its life cycle.

2. The discovery of penicillin has recently been taken up by Gwyn MacFarlane in *Alexander Fleming: The Man and the Myth* (Boston, 1984).

3. Hamilton Smith, Daniel Nathans, and Werner Arber shared a Nobel Prize in 1978 for work related to the nature and uses of restriction endonucleases. See J. Cherfas, *Man Made Life: An Overview of the Science, Technology and Commerce of Genetic Engineering* (New York, 1983), 42.

4. W. T. Schrader and B. W. O'Malley (eds.), *Laboratory Methods Manual for Hormone Action and Molecular Endocrinology* (5th ed.; Houston, Tex., 1980); D. T. Denhardt, "A Membrane-Filter Technique for the Detection of Complementary DNA," *Biochemical Biophysical Research Communications,* 23 (1966), 641–46; M. Chase, "Walking Tall: New Growth Hormone Can Benefit Children Physically and Socially," *Wall Street Journal,* June 24, 1983, p. 1. Perhaps the best compendium of recent advances in recombinant DNA technology is in *Science,* 222 (1983). This volume, completely dedicated to cDNA technology, discusses the implications of this research for immunology, genetics, and other disciplines. On the cover is the classic picture of litter-mate rats, one markedly larger than the other as a result of the introduction of a human growth hormone gene.

5. R. Lewin, "The Birth of Recombinant RNA Technology: A Newly Developed Technique Promises to Allow the Production of Large Quantities of Any Chosen RNA by Hooking It to a Replication Vehicle," *Science,* 222 (1983), 1313–15.

6. A. M. Maxam and W. Gilbert, "A New Method for Sequencing DNA," *PNAS,* 74 (1977), 560–64; F. Sanger, "Determination of Nucleotide Sequences in DNA," *Science,* 214 (1981), 1205–1210. Sanger and Gilbert shared a Nobel Prize with Paul Berg in 1980.

7. The DNA sequencing method of Maxam and Gilbert is conceptually similar to that of Sanger. Once a landmark is formed in the DNA strands to be sequenced with a restriction nuclease, the ends of the strands are radioactively labeled with ^{32}P. Next, the DNA is run on a denaturing electrophoretic gel, which separates the two strands of DNA and generates single strands. The single-stranded DNA is then divided into four batches, each of which is treated with a specific reagent. Batch 1 is treated with dimethyl sulfate, which adds a methyl group to purines and particularly to guanine. The second batch is treated with acid, which nonselectively attacks purines. It removes guanines and adenines from the DNA but does not interrupt the sugar backbone. The third batch is treated with hydrazine, which opens the ring of pyrimidines, cytosine, and thymine. In the presence of salt, hydrazine pref-

erentially attacks cytosine, and this is the circumstance in which the fourth batch of DNA is incubated. If the proper concentrations of dimethyl sulfate, acid, and hydrazine are added to the four batches, on average each guanine will be modified in one chain or another in batch 1, each purine (adenine or guanine) in one or another chain in batch 2, each pyrimidine (cytosine or thymine) in batch 3, and each cytosine in one or another chain in batch 4. Next, all batches are treated with piperazine, which breaks the DNA at each base that was modified in the above four reactions. The resultant products from each batch are then subjected to electrophoresis to divide the generated fragments by size. When the resulting gels are subjected to autoradiography, the gel derived from the first batch will reveal the positions of all guanines, those derived from the second will reveal the guanines and adenines, those of the third the cytosine and thymine, and those of the fourth the cytosine. From these gels the DNA sequence is easily determined.

8. E. S. McAuliff and K. McAuliff, "The Genetic Assault on Cancer," *New York Times Magazine,* October 24, 1982, pp. 39–54; R. Watt *et al.,* "Nucleotide Sequence of Cloned cDNA of Human C-myc Oncogene," *Nature,* 303 (1983), 725–28; H. Garoff *et al.,* "Nucleotide Sequence of cDNA Coding for Semliki Forest Virus Membrane Glycoproteins," *Nature,* 288 (1980), 236–41 (contains excellent examples of endonuclease maps and sequencing gels); K. E. Mostov *et al.,* "The Receptor for Transepithelial Transport of IgA and IgM Contains Multiple Immunoglobulin-like Domains," *Nature,* 308 (1984), 37–43 (describes the novel utilization of DNA sequence analysis to demonstrate that the gut receptor for immunoglobulin shares a common structure with the immunoglobulins themselves; thus, the technique of sequence analysis sheds light on the evolutionary development and basic nature of control mechanisms). See also A. F. Williams, "The Immunoglobulin Super Family Takes Shape," *Nature,* 308 (1984), 12–13; A. S. Whitehead *et al.,* "Isolation of Human C-Reactive Protein Complementary DNA and Localization of the Gene to Chromosome 1," *Science,* 221 (1983), 69–71; R. G. Elles *et al.,* "Absence of Maternal Contamination of Chorionic Villi Used for Fetal-Gene Analysis," *NEJM,* 308 (1983), 1433–35 (describes the use of villous biopsy and hybridization analysis for the very early diagnosis of genetic disease *in utero*); P. Newmark, "Molecular Diagnostic Medicine," *Nature,* 307 (1984), 11; "Capsule Comment: Antenatal Sex Determination While-You-Wait," *Hospital Practice* (March, 1984), 31 (describes a rapid screening test for antenatal sex determination, used for the antenatal diagnosis of sex-linked disorders, such as Duchenne-type muscular dystrophy, hemophilia, and Leisch-Nyhan syndrome, where fetal sex is determined first because further testing is usually not necessary if the fetus is female); P. Leder, "Mechanisms of Gene Evolution," *JAMA,* 248 (1982), 1582–91 (describes what is currently known about genetic mechanisms, including jumping genes, and relates this information to evolution); P. Echeverria *et al.,* "Identification by DNA Hybridization of Enterotoxigenic *Escherichia coli* in Homes of Children with Diarrhea," *Lancet,* 1 (1984), 63–65 (demonstrates the utility of hybridization techniques for the diagnosis of infectious diseases); J. E. Godine *et al.,* "α-Subunit of Rat Pituitary Glycoprotein Hormones: Primary Struc-

ture of the Precursor Determined from the Nucleotide Sequence of Cloned cDNAs," *JBC*, 257 (1982), 8368–71; R. Patient, "DNA Hybridization—Beware," *Nature*, 308 (1984), 15–16 (describes some pitfalls in interpreting hybridization data and extrapolating it to conclusions about cell activity); R. F. Doolittle *et al.*, "Computer-Based Characterization of Epidermal Growth Factor Precursor," *Nature*, 307 (1984), 558–60 (describes the dynamic nature of the evolution of the epidermal growth factor gene and utilizes computer analysis of genetic sequences to unravel the mysteries of evolution); K. H. Cowan *et al.*, "Dihydrofolate Reductase Gene Amplification and Possible Rearrangement in Estrogen-Responsive Methotrexate-Resistant Human Breast Cancer Cells," *JBC*, 257 (1982), 15079–86.

9. The initial observation of exons was actually made by hybridizing mRNA to genomic DNA. Loops were detected by electronmicroscopy. See Cherfas, *Man Made Life*, 218–76, for a historical account of this discovery. See also Y. Nabeshima *et al.*, "Alternative Transcription and Two Modes of Splicing Result in Two Different Light Chains from One Gene," *Nature*, 308 (1984), 333–38 (another example of message processing resulting in the synthesis of multiple proteins from the same gene); C. S. Craik *et al.*, "Intron-Exon Splice Junctions Map at Protein Surfaces," *Nature*, 299 (1982), 180–82 (raises the possibility that the functionally important portions of proteins, namely the surfaces, occur at intron-exon spliced junctions, suggesting that alterations in splicing site could lead to evolutionary development of proteins of new function); Walter Gilbert, "Why Genes in Pieces," *Nature*, 271 (1978), 501; K. M. Lang and R. A. Spritz, "RNA Splice Site Selection: Evidence for a 5′–3′ Scanning Model," *Science*, 220 (1983), 1351–55; M. G. Rosenfeld, S. G. Amara, and R. N. Evans, "Alternative RNA Processing: Determining Neuronal Phenotype," *Science*, 225 (1984), 1315–20.

Chapter VI

1. J. L. Marx, "Diabetes—A Possible Autoimmune Disease: Insulin-Dependent Diabetes May Be Caused by an Immune Attack on Islet Cells, But Efforts to Prevent It with Immunosuppression Are Still Controversial," *Science*, 225 (1984), 1381–83.

2. A. Joyner *et al.*, "Retrovirus Transfer of a Bacterial Gene into Mouse Haematopoietic Progenitor Cells," *Nature*, 305 (1983), 556–58. This work demonstrates the feasibility of using retroviruses as a vehicle for inserting new genes into specific body cells. As such, it makes more palpable the emerging reality of genetic therapy. An exciting area of investigation is the determination of control signals for DNA transcription and replication. For example, consider the work of Y. Gruenbaum *et al.*, "Methylation of Replicating and Post-Replicated Mouse L-Cell DNA," *PNAS*, 80 (1983), 4919–21. The thoughtful reader will realize that the engineered cells in this example must not only *synthesize* insulin appropriately, but also must *secrete* it into the blood. Thus, the bioengineer must either induce secretory activity along with a capacity for insulin synthesis or must make use of cells in the initial biopsy that possess either constitutive or regulated secretory pathways. Fortunately, such cells exist.

Chapter VII

1. There are several mechanisms by which stroke can be produced. First is thrombotic vascular occlusion as described in the text. An embolic stroke occurs when a small clot (often produced over an area of vascular damage) or other material travels to a cerebral vessel and blocks it. Finally, stroke can be produced if a diseased vessel ruptures, resulting in bleeding into the brain. But in the great majority of strokes, as in the great majority of heart attacks, vascular disease, and more particularly atherosclerotic vascular disease, plays a critical role. It is in these instances that the common origin of stroke and heart attack is clear. H. Ohnishi *et al.*, "A New Approach to the Treatment of Atherosclerosis and Trapidil as an Antagonist to Platelet-Derived Growth Factor," *Life Sciences*, 28 (1981), 1641–46. This paper, in my view, is a landmark in medicine in that it heralds the beginning of the cellular biological attack on atherogenesis. A drug is described that has the capacity of blocking the action of platelet-derived growth factor on the arterial smooth muscle cell and therefore prevents the development of atherosclerosis in experimental models. Whether the particular drug involved is safe and effective in man is unclear, but the approach heralds a new era in the prevention of heart disease and stroke. See also R. Ross *et al.*, "A Platelet-Dependent Serum Factor that Stimulates the Proliferation of Arterial Smooth Muscle Cells *in Vitro*," *PNAS*, 71 (1974), 1207–1210; R. Weinstein and M. B. Stemerman, "Hormonal Requirements for Growth of Arterial Smooth Muscle Cells *in Vitro*: An Endocrine Approach to Atherosclerosis," *Science*, 212 (1981), 818–20.

2. *Internal Medicine News*, December 1–14, 1984, p. 3. Reduction of cardiovascular mortality has resulted in a 40 percent decline in age-adjusted mortality in the last three decades. See also *Medical Tribune*, March 21, 1984, p. 2, which discusses research in Japan and San Francisco that further indicates that calcium channel blockers at high dose seem to retard atherogenesis; and R. Ross, "The Pathogenesis of Atherosclerosis—An Update," *NEJM*, 314 (1986), 488–97.

3. For those interested in gaining some insight into just how complex this cellular biology of the arterial wall is, see R. N. Re *et al.*, "Renin Synthesis by Canine Aortic Smooth Muscle Cells in Culture," *Life Sciences*, 30 (1981), 99–106; R. N. Re *et al.*, "Nuclear-Hormone Mediated Changes in Chromatin Solubility," *Biochemical Biophysical Research Communications*, 110 (1982), 61–68; and R. N. Re and M. Parab, "Effect of Angiotensin II on RNA Synthesis by Isolated Nuclei," *Life Sciences*, 34 (1984), 647–51. These papers demonstrate not only that one need consider tissue hormone systems, such as the renin system in the arterial wall, but also potentially intracellular hormonal systems such as the intracellular renin-angiotensin system. For a review, see R. N. Re, "The Cellular Biology of the Renin-Angiotensin Systems," *AIM*, 144 (1984), 2037–2041; J. D. Bagdade and P. V. Subbaiah, "Atherosclerosis and Oral Contraceptive Use: Serum from Oral Contraceptive Users Stimulates Growth of Arterial Smooth Muscle Cells," *Atherosclerosis*, 2 (1982), 170–76 (describes the use of tissue culture techniques to track down serum factors associated with atherosclerosis, such as that seen in some people following

oral contraceptive use); J. Davignon, R. Dufour, and M. Cantin, "Atherosclerosis and Hypertension," in J. Genest *et al.* (eds.), *Hypertension: Physiopathology and Treatment* (2nd ed.; New York, 1983), 810–52; S. Fowler *et al.*, "Subcellular Fractionation and Morphology of Calf Aortic Smooth Muscle Cells: Studies on Whole Aorta, Aortic Explants, and Subcultures Grown Under Different Conditions," *Journal of Cellular Biology,* 75 (1977), 166–84 (describes the cell culture models used in the study of atherogenesis and illustrates how the detailed study of cells in culture can provide information that potentially can be extrapolated back to the intact animal); *Internal Medicine News,* March 15–31, 1984, p. 37 (describes the possibility that viral infections or other events capable of initiating immune complex formation in the blood could lead to endothelial damage and atherogenesis).

4. J. L. Marx, "Oncogenes Amplified in Cancer Cells," *Science,* 223 (1984), 40–41; Ohnishi *et al.*, "A New Approach to the Treatment of Atherosclerosis," 1641–46; P. Newmark, "Cancer Genes—Processed Genes—Jumping Genes," *Nature,* 296 (1982), 393–94; C. J. Marshall, "A Transforming Gene Present in Human Sarcoma Cell Line," *Nature,* 299 (1982), 171–73; R. Dalla-Favera *et al.*, "Human *c-myc onc* Gene Is Located on the Region of Chromosome 8 That Is Translocated in Burkitt Lymphoma Cells," *PNAS,* 79 (1982), 7824–28; H. C. Pitot, "Chemicals in Cancer: Initiations and Promotion," *Hospital Practice* (July, 1983), 101–113 (discusses the role of onc genes in the multistep process that leads to cancer); A. P. Albino, "Transforming *ras* Genes from Human Melanoma: A Manifestation of Tumour Heterogeneity," *Nature,* 308 (1984), 69–72 (demonstrates that not all tumor cells in a given melanoma need have the same onc gene activated—in this tumor line apparently the *ras* gene is coincidentally activated and is not responsible for the transformation); P. M. Boffey, "U.S. and British Researchers Find Cancer a 2-Step Process," New York *Times,* August 18, 1983, p. 1; H. Land *et al.*, "Tumorogenic Conversion of Primary Embryo Fibroblasts Requires at Least Two Cooperating Oncogenes," *Nature,* 304 (1983), 596–602 (demonstrates the need for multiple oncogene interactions for the production of cancer in normal cells and describes the techniques used in the search for oncogenes); R. F. Newbold and R. W. Overell, "Fibroblast Immortality Is a Prerequisite for Transformation by *EJ c-Ha-ras* Oncogene," *Nature,* 304 (1983), 648–51; C. McGill, "Advances in Genetics Stir Cancer Scientists," New York *Times,* July 24, 1983, p. 34; T. F. Devel and J. S. Huang, "Platelet-Derived Growth Factor: Structure, Function, and Roles in Normal and Transformed Cells," *JCI,* 74 (1984), 669–76; A. Balmain and I. B. Pragnell, "Mouse Skin Carcinomas Induced *in Vivo* by Chemical Carcinogens Have a Transforming Harvey-ras Oncogene," *Nature,* 305 (1983), 72–74 (confirms that the oncogene mechanism is involved in the production of environmentally induced cancers as well as in "spontaneous" and virally induced neoplasms); R. Watt *et al.*, "Nucleotide Sequence of Cloned cDNA of Human C-myc Oncogene," *Nature,* 303 (1983), 725–28; R. C. Gallo, "The Virus-Cancer Story," *Hospital Practice* (June, 1983), 79–89; J. L. Marx, "*onc* Gene Related to Growth Factor Gene: The Finding Supports the View that an *onc* Gene May Make Cells Cancerous by Inappropriately Producing a Substance that Regulates Normal

Growth," *Science,* 221 (1983), 248; R. C. Gallo and F. Wong-Staal, "Retroviruses as Etiological Agents of Some Animal and Human Leukemias and Lymphomas and as Tools for Elucidating the Molecular Mechanisms of Leukemogenesis," *Blood,* 60 (1982), 545–57; R. C. Gallo and F. Wong-Staal, "Human T-Cell Leukemia-Lymphoma Virus (HTLV) and Human Viral *onc* Gene Homologues," in T. E. O'Conner and F. J. Rauscher, Jr. (eds.), *Oncogenes and Retroviruses: Evaluation of Basic Findings and Clinical Potential* (New York, 1983), 223–42.

5. "Worries About Infectious Cancer," *Nature,* 302 (1983), 567; D. W. Blayney *et al.,* "The Human T-Cell Leukemia-Lymphoma Virus in the Southeastern United States," *JAMA,* 250 (1983), 1048–1052; W. A. Blattner *et al.,* "Human T-Cell Leukemia-Lymphoma Virus and Adult T-Cell Leukemia," *JAMA,* 250 (1983), 1074–1084. An interesting update on the HTLV story is found in *Hospital Practice* (March, 1984), 29, which describes the transmission of T-cell leukemia virus to non–T cells and suggests that HTLV may be a misnomer. The virus actually may cause osteogenic sarcoma and perhaps other sarcomas. As the authors of this report state, "reverse transcriptase is a powerful tool, and the virus may have more to gain by reproducing itself than by transforming its host cell into a tumor that destroys that host. The model of hepatitis B virus as a cause of cellular destruction (hepatitis), prolonged viral proliferation (persistent hepatitis), or hepatoma illustrates the issue. AIDS, which has in fact just recently been found to have an association with HTLV, may represent the whole ominous spectrum of the above possibilities—from immunosuppression to Kaposi's sarcoma." The actual work describing the transmission of the virus to a nonlymphoid human osteogenic sarcoma cell line is reported by P. Clapham *et al.,* "Productive Infection and Cell-Free Transmission of Human T-Cell Leukemia Virus in a Nonlymphoid Cell Line," *Science,* 222 (1983), 1125. See also Barre Sinoussi *et al.,* "Isolation of a T-Lymphocyte Retrovirus from a Patient at Risk for Acquired Immune Deficiency Syndrome (AIDS)," *Science,* 220 (1983), 868–71; J. A. Levy *et al.,* "Isolation of Lymphocytopathic Retroviruses from San Francisco Patients," *Science,* 225 (1984), 840–42; S. F. Josephs *et al.,* "Human Proto-Oncogene Nucleotide Sequences Corresponding to the Transforming Region of Simian Sarcoma Virus," *Science,* 223 (1984), 487–91; N. Yamomoto *et al.,* "Unique Cell Lines Harboring Both Epstein-Barr Virus and Adult T-Cell Leukemia Virus Established from Leukemia Patients," *Nature,* 299 (1982), 367–69; R. F. Doolittle *et al.,* "Simian Sarcoma Virus *onc* Gene, *V-sis,* Is Derived from the Gene (or Genes) Encoding a Platelet-Derived Growth Factor," *Science,* 221 (1983), 275–76. For an update report demonstrating the relationship between PDGF and the three onc-genes with which it is related functionally, see W. Kruijer *et al.,* "Platelet-Derived Growth Factor Induces Rapid But Transient Expression of the *C-fos* Gene and Protein," *Nature,* 312 (1984), 711–16; and Rolf Muller, "Induction of C-fos Gene and Protein by Growth Factors Precedes Activation of *C-myc,*" *Nature,* 312 (1984), 716–20. It seems PDGF, itself related to the *C-sis* onc gene, acts on normal cells to cause growth by stimulating the *C-fos* and *C myc* onc genes. R. Weiss, "Oncogenes and Growth Factors,"

Nature, 304 (1983), 1219; F. Ruscetti *et al.,* "The Interaction of Human T-Cell Growth Factor with Normal and Neoplastic Cells," in E. Mihich (ed.), *Biological Responses in Cancer: Progress Toward Potential Applications* (New York, 1982), 20–28 (points out that growth factors play an important role in the immune system just as they do in cancer development and atherogenesis); New York *Times,* July 3, 1983. Sec. E, p. 14.

6. R. Trubo, "Genetic Manipulation with Retroviruses May Lead to Lesch-Nyhan Treatment," *Medical World News,* October 24, 1983, p. 9 (describes ongoing efforts to make genetic therapy in man a clinical reality); S. Budransky, "Gene Therapy: Quick Fix Not in the Cards," *Nature,* 306 (1983), 414. For an excellent review of the potential of genetic engineering for clinical medicine, including a discussion of the possible use of retroviruses as therapeutic agents, see "Medical News," *JAMA,* 253 (1985), 13–18.

7. S. K. Arya *et al.,* "T-Cell Growth Factor Gene: Lack of Expression in Human T-Cell Leukemia-Lymphoma Virus-Infected Cells," *Science,* 223 (1984), 1086–1087 (suggests that, although some oncogenes may code for growth factors, others may in fact code for abnormal cell receptors for growth factors; possibly these abnormal receptors are continually turned on, mimicking thereby the presence of growth factors—this is currently a very controversial area of theory and research); J. L. Collins *et al.,* "*In Vitro* Surveillance of Tumorigenic Cells Transformed *in Vitro,*" *Nature,* 299 (1982), 169–71.

8. D. A. Galloway and J. K. McDougall, "The Oncogenic Potential of *Herpes Simplex* Viruses: Evidence for a 'Hit-and-Run' Mechanism," *Nature,* 302 (1983), 21–24.

9. A. E. Friedman-Kien *et al.,* "Kaposi's Sarcoma and Pneumocystis Pneumonia in New York and California," *Morbidity and Mortality Weekly Report,* 30 (1981), 305–308; M. S. Gottlieb *et al.,* "*Pneumocystis Carinii* Pneumonia and Mucosal Candidiasis in Previously Healthy Homosexual Men," *NEJM,* 305 (1981), 1425–31; D. T. Durack, "Opportunistic Infections and Kaposi's Sarcoma in Homosexual Men," *NEJM,* 305 (1981), 1465–67; A. J. Ammann *et al.,* "Acquired Immunodeficiency in an Infant: Possible Transmission by Means of Blood Products," *Lancet,* 1 (1983), 956–58; M. Essex *et al.,* "Antibodies to Cell Membrane Antigens Associated with Human T-Cell Leukemia Virus in Patients with AIDS," *Science,* 220 (1983), 859–62; H. C. Lane *et al.,* "Abnormalities of B-Cell Activation and Immunoregulation in Patients with Acquired Immunodeficiency Syndrome," *NEJM,* 309 (1983), 453–58; A. J. Ammann *et al.,* "B-Cell Immunodeficiency in Acquired Immune Deficiency Syndrome—An Acquired Immune Deficiency Syndrome," *JAMA,* 251 (1984), 1447–49 (demonstrates abnormalities in the lymphocytes that produce antibodies as well as in the delayed hypersensitivity-related lymphocytes in patients with AIDS—as such, it points up the relationships between the various components of the cellular and humoral immune systems); J. E. Groopman and P. A. Volberding, "The AIDS Epidemic: Continental Drift," *Nature,* 306 (1984), 211–12; H. A. Jaffe *et al.,* "Transfusion-Associated AIDS:

Serological Evidence of Human T-Cell Leukemia Virus Infection of Donors," *Science*, 223 (1984), 1309–11; J. Sonnabend *et al.*, "Acquired Immunodeficiency Syndrome, Opportunistic Infections, and Malignancies in Male Homosexuals: A Hypothesis of Etiological Factors in Pathogenesis," *JAMA*, 249 (1983), 2370–74. This article suggests that "profound promiscuity" could play a role in the genesis of immune deficiency syndrome. The article was answered by a letter from H. H. Handsfield (*JAMA*, 251 [1984], 341) in which the argument was raised that perhaps multiple factors could predispose to infection with the causative agent of AIDS. Thus repeated exposure to semen, cytomegalovirus, or other agents may cause, in this view, a subclinical immunodeficiency that is nonprogressive but permits infection by an AIDS-specific virus, which then results in irreversible immunosuppression. However, D. J. McShane and N. R. Schram argue (*JAMA*, 251 [1984], 341) that the disease likely can be transmitted simply through blood as in the transmission to hemophiliacs by blood products as well as from parent to child. These physicians suggest that any theory basing AIDS on assumed "profound promiscuity" fails to explain all facts available to date and could potentially be dangerous in that it would suggest to those not "profoundly promiscuous" that they are not at risk.

10. H. Mitsuya *et al.*, "Suramin Protection of T-Cells *in Vitro* Against Infectivity and Cytopathic Effect of HTLV-III," *Science*, 226 (1984), 172–74.

11. P. Newmark, "Molecular Diagnostic Medicine," *Nature*, 307 (1984), 11–12.

12. S. M. Larson *et al.*, "Diagnostic Imaging of Malignant Melanoma with Radiolabeled Anti-Tumor Antibodies," *JAMA*, 249 (1983), 811–12; J. Q. Trojanowski and V. M.-Y. Lee, "Anti-Neurofilament Monoclonal Antibodies: Reagents for the Evaluation of Human Neoplasmas," *AN*, 59 (1983), 155–58; K. Naito *et al.*, "Autologous and Allogeneic Typing of Human Leukemia Cells: Definition of Surface Antigens Restricted to Lymphocytic Leukemia Cells," *PNAS*, 80 (1983), 2341–45; S. M. Larson, "Localization of ^{131}I-Labeled $_p$97-Specific Fab Fragments in Human Melanoma as a Basis for Radiotherapy," *JCI*, 72 (1983), 2101–2114; R. A. Miller *et al.*, "Treatment of B-Cell Lymphoma with Monoclonal Anti-Idiotype Antibody," *NEJM*, 306 (1982), 517–22.

13. N. Kitamura, "A Single Gene for Bovine High Molecular Weight and Low Molecular Weight Kininogens," *Nature*, 305 (1983), 545–49 (demonstrates how cDNA techniques can be utilized to elucidate the relationship between different molecular forms of serum proteins and hormones); P. A. Johnson and A. H. Rossof, "The Role of the Human Tumor Stem Cell in Medical Oncology," *AIM*, 143 (1983), 111–13; R. C. Dodd *et al.*, "Vitamin D Metabolites Change the Phenotype of Monoblastic U937 Cells," *PNAS*, 80 (1983), 7538–41; K. McAuliff and E. S. McAuliff, "Keeping Up with the Genetic Revolution," *New York Times Magazine*, November 6, 1983, pp. 41–97; M. Chase, "Walking Tall: New Growth Hormone Can Benefit Children Physically and Socially," *Wall Street Journal*, June 24, 1983, p. 1; E. I. Simpson *et al.*, "Absence of Parathyroid Hormone Messenger

Note to Page 102

RNA in Nonparathyroid Tumors Associated with Hypercalcemia," *NEJM*, 309
(1983), 325–30 (demonstrates how hybridization techniques can be used to answer
basic questions in medicine, such as whether or not tumors cause hypercalcemia by
the production of a hormone normally found only in the parathyroid glands—apparently tumors do not usually produce the message for this hormone, and therefore the mechanism of the hypercalcemia they cause must be sought elsewhere);
"Damon Biotech Micro-Encapsulated Pancreas Cells Produce Insulin," *Blue Sheet*,
April 22, 1983, pp. P, R–9; "Antiviral Vaccines/Cancer Prophylactics?" *JAMA*,
250 (1983), 15–20; G. Kolata, "Globin and Gene Studies Create Puzzles," *Science*, 223 (1984), 470–71 (describes the use of 5-azacytidine, which was felt to be
capable of altering the methylation status of specific genes and thereby potentially
turning on helpful genes, in patients with deficient production of normal hemoglobin—the drug seems to help some patients at least temporarily and to demethy-
late certain genes, but it is likely that the two effects are unrelated, leading to the
significant question of how this molecular biological attack on inherited anemia
actually works); T. Madrup-Poulen *et al.*, "DNA Sequences Flanking the Insulin
Gene on Chromosome 11 Confer Risk of Atherosclerosis," *Lancet*, 1 (1984),
250–53 (describes the use of recombinant DNA techniques to detect genomic arrangements that predispose to atherosclerosis—this is the first step in attacking not
only the cellular but the molecular-genetic basis of atherosclerosis); M. Kozak,
"Point Mutations Close to the AUG Initiator Code Can Affect the Efficiency of
Translation of Rat Preproinsulin *in Vivo*," *Nature*, 308 (1984), 241–47 (demonstrates that minor genetic abnormalities can potentially play a role in determining
the level of protein hormone secretion in various states and therefore possibly in the
genesis of diseases such as noninsulin-dependent diabetes mellitus); Y. Shing *et
al.*, "Heparin Affinity: Purification of a Tumor-Derived Capillary Endothelial Cell
Growth Factor," *Science*, 223 (1984), 1296–98 (discusses another novel cellular
biological attack on cancer that hinges on the fact that cancers must produce a factor to induce the ingrowth of new vessels if the cancers are not to die of starvation at
a very small size—from the study of this growth factor may come an effective
means of limiting the growth of some cancers); "Apolipoprotein Gene Abnormalities Tied to Severe Atherosclerosis," *Internal Medicine News*, March 15–31,
1984, p. 14 (describes the work of Dr. Jan L. Breslow at Rockefeller University,
New York, on the genetic basis of cholesterol abnormalities—it appears that the
specific genes responsible for the production of abnormal cholesterol binding proteins in blood and hence of cholesterol abnormalities have been identified and that
molecular genetics is thus beginning to zero in on one of the major risk factors for
heart disease); "Sees Wider Use of Marker for Huntington's in '85," *Internal Medicine News*, March 15–31, 1984, p. 1 (describes the clinical utility of the newly
devised hybridization method for the diagnosis of Huntington's chorea, a degenerative nerve disorder); "A Postwar State of MS in Faroes," *Medical Tribune*, March
21, 1984, p. 2 (describes the fact that the Nordic population of the Faroes experienced an extremely low prevalence of multiple sclerosis until immediately follow-

ing World War II, when the only obvious societal change that occurred was the British occupation, leading one to believe that an infectious agent was introduced into the population); M. B. Sporn, "Retinoids and Suppression of Carcinogenesis," *Hospital Practice* (October, 1983), 83–98 (reviews the use of retinoids [vitamin A analogues] in the possible prevention of cancer—more recent work suggests an effect of vitamin A on *onc* gene transcription); D. Jahner *et al.*, "*De Novo* Methylation and Expression of Retroviral Genomes During Mouse Embryogenesis," *Nature,* 298 (1982), 623–28. A more recent example of the potential of Bioburst to aid in the attack on cancer is the cloning of the gene for so-called lymphotoxin, a lymphocyte product capable of killing, by an as-yet unknown mechanism, a variety of types of tumor cells. This achievement should soon make this material available in quantity for the therapy of human cancer. For details, see P. W. Gray *et al.*, "Cloning and Expression of cDNA for Human Lymphotoxin, A Lymphokine with Tumor Necrosis Activity," *Nature,* 312 (1984), 721–24; and D. Pennica *et al.*, "Human Tumour Necrosis Factor: Precursor, Structure, Expression, and Homology to Lymphotoxin," *Nature,* 312 (1984), 724–29. Some other applications of Bioburst technology are described in S. A. Rosenberg *et al.*, "Observations on the Systemic Administration of Autologous Lymphokine-Activated Killer Cells and Recombinant Interleukin-2 to Patients with Metastatic Cancer," *NEJM,* 313 (1985), 1435–92; M. Davis *et al.*, "Chromosome Translocation Can Occur on Either Side of the *c-myc* Oncogene in Burkitt's Lymphoma Cells," *Nature,* 308 (1984), 286–88; M. Schwab *et al.*, "Chromosome Localization in Normal Human Cells and Neuroblastomas of a Gene Related to *c-myc,*" *Nature,* 308 (1984), 288–91; S. Rasheed *et al.*, "Nucleotide Sequence of the Rasheed Rat Sarcoma Virus Oncogene: New Mutations," *Science,* 221 (1983), 155–57; T. G. Krontiris, "The Emerging Genetics of Human Cancer," *NEJM,* 309 (1983), 404–409; D. J. Capon *et al.*, "Complete Nucleotide Sequence of the T24 Human Bladder Carcinoma Oncogene and Its Normal Homologue," *Nature,* 302 (1983), 33–37; G. Winchester, "p53 Protein and Control of Growth," *Nature,* 303 (1983), 660–61; D. C. Dean *et al.*, "A 5'-Flanking Sequence Essential for Progesterone Regulation of an Ovalbumin Fusion Gene," *Nature,* 305 (1983), 551–54 (describes the use of cDNA technology for elucidating the basic mechanisms by which hormones and drugs work in the human body—as such, it portends an entirely new approach to pharmacotherapy.).

Chapter VIII

1. J. Rifkin and N. Perlas, *Algeny* (New York, 1983), 247–55; C. Normin, "Clerics Urge Ban on Altering Germ Line Cells," *Science,* 220 (1983), 1360–61.

2. S. J. O'Brian *et al.*, "The Cheetah Is the Depauperate in Genetic Variation," *Science,* 221 (1983), 459–61; A. G. Motulsky, "Impact of Genetic Manipulation on Society and Medicine," *Science,* 219 (1983), 135–40 (a discussion of past and present human activities with genetic implications); S. W. Mosher, "Why Are Baby Girls Being Killed in China?" *Wall Street Journal,* July 25, 1983, p. 11; "Singapore Backs Babies for Elite: Better-Educated People Are Urged to Multiply,"

New York *Times,* February 12, 1984, p. 17 (presents one of the more traditional, albeit primitive, modes of genetic engineering).

3. Stephen J. Gould, *Hens' Teeth and Horses' Toes: Further Reflections in Natural History* (New York, 1983), 335.

4. R. D. Palmiter *et al.,* "Dramatic Growth of Mice That Developed from Eggs Micro-Injected with Metallophionein–Growth Hormone Fusion Genes," *Nature,* 300 (1982), 611–15 (demonstrates that the introduction of foreign genes and their control elements into embryos permits man to alter animals or entire species genetically; in a very real sense, it lays the groundwork for practical germ line engineering).

5. "Filly Born of Frozen Embryo," New York *Times,* June 26, 1983, p. 11; "Australian Scientists Have Reported the First Successful Pregnancy Using a Frozen *in Vitro* Fertilized Embryo," *Physicians Washington Report* (June, 1983), 1 (describes work at Monash University, Melborne); P. Lutjen *et al.,* "The Establishment and Maintenance of Pregnancy Using *in Vitro* Fertilization and Embryo Donation in a Patient with Premature Ovarian Failure," *Nature,* 307 (1984), 174–75. This paper describes the application in humans of techniques previously used only in animals. Taken together with recent discoveries that fertilized ova can be frozen for long periods of time and that fertilized animal ova can be quartered, with each segment producing an entire animal after implantation into a mother's womb, this study presages a form of genetic engineering with practical implications for human reproduction. It is not impossible that fertilized ova can be cloned, in the sense that from one ova several genetically identical children could be produced. See also L. McLaughlin, "Frozen Embryos," Boston *Globe,* April 16, 1984, p. 48 (describes the birth in Australia of the baby Zoë, whose first months of existence were spent frozen in liquid nitrogen at −325°F; on page 45 of the same paper are pictures of Zoë at the five-cell stage of development and at two weeks after birth); H. Lancaster, "Firm Offering Human-Embryo Transfers for Profit Stirs Legal and Ethical Debates," *Wall Street Journal,* March 7, 1984, p. 33; J. F. Henahan, "Fertilization, Embryo Transfer Procedures Raise Many Questions," *JAMA,* 252 (1984), 877–82; C. B. Fehilly *et al.,* "Interspecific Chimaerism Between Sheep and Goat," *Nature,* 306 (1983), 634–36; S. Meinecke-Tillmanns and B. Meinecke, "Experimental Chimaeras—Removal of Reproductive Barrier Between Sheep and Goat," *Nature,* 307 (1984), 636–38. The last two papers demonstrate that germ line engineering of sorts can be carried out even without recombinant DNA technology. The skillful manipulation of early-stage embryos permits the development of chimaeras as if by engineering. Also of note in this regard is the recent report in the New Orleans *Times Picayune* (March 6, 1984) of experimental work at Louisiana State University, in which a fertilized ovum was divided into four smaller parts, each of which was then reimplanted into a cow. Two sets of identical twin animals were then born from two mothers. This experiment offers the possibility of cloning intact animals and will have major implications for the breeders of animals.

6. "Last Gape: The Tasmanian Tiger," *Nature,* 307 (1984), 411. Pictured here is the last characteristic open-mouth gape (jaw open to greater than 120°) of the

Tasmanian tiger, an animal that became extinct in 1936. A genetic safety net could have saved at least the seeds of this species.

7. C. Austin, "Conferees Weigh Work with Genes: Scientists and Theologians in Accord on Putting Stress on Cures for Disease," New York *Times,* August 7, 1983, p. 27.

Chapter IX

1. J. Cherfas, *Man Made Life: An Overview of the Science, Technology and Commerce of Genetic Engineering* (New York, 1983), 130–31.

2. T. Beardsley, "Biotechnology: Low Risks at a Premium," *Nature,* 308 (1984), 579 (describes "biotech protection," an insurance package geared to the needs of biotechnical industries; "loss engineers" will supervise safety practices for the insurance company and make on-site recommendations for improvements, thus providing another layer of oversight); C. Earl, "Rules for Freed Organisms Planned," *Nature,* 306 (1983), 5; M. Sun, "Congress Ponders rDNA and Environmental Risks," *Science,* 221 (1983), 136–37; "Genetic Engineering Raises Concerns of Senator Hatfield (R-Oregon)," *Blue Sheets,* June 15, 1983, p. 8; P. M. Boffey, "Can Those Skirting Nature Be Kept in Line?" New York *Times,* February 12, 1984, Section E, p. 20. This last article summarizes the views of the scientific community as well as those of the ecological opposition. Dr. Matthew S. Meselson, a leading scholar of biological warfare, notes that there would be little advantage to using molecular biology to create new bacteriological warfare agents, and Alexander M. Capron, executive director of a presidential commission that studied the ethical issues raised by biomedical advances, says that "the commission found that the things now in prospect are nothing for people to get concerned or alarmed about." The difficult issues, he thinks, are far in the future.

3. W. Kucewicz, "Word from Behind the Iron Curtain: Beyond 'Yellow Rain'—The Threat of Soviet Genetic Engineering, An Update," *Wall Street Journal,* December 28, 1984, pp. 1, 9.

4. C. Norman, "Genetically Engineered Plants Get a Green Light," *Science,* 222 (1983), 35 (the release of genetically engineered organisms has since been halted pending further study); C. Earl, "Biotechnology Regulation: Rules for Freed Organisms Planned," *Nature,* 306 (1983), 5 (describes the legal opinion that freed organisms should be controlled under the Federal Toxic Substances Control Act [TOSCA] and that the release of ice-nucleation bacteria, which normally cause frost sensitivity on plants, whose ice-nucleation genes were deleted by recombinant DNA techniques should come under the jurisdiction of the Federal Insecticide, Fungicide, and Rodenticide Act [FIFRA]); S. D. Holmberg, J. G. Wells, and M. L. Cohen, "Animal-to-Man Transmission of Antimicrobial-Persistent Salmonella: Investigations of U.S. Outbreaks, 1971–1983," *Science,* 225 (1984), 833–35 (this report warns that antibiotics given to animals can produce antibiotic-resistant salmonella, which can then infect man, and suggests that we cannot be too sanguine about the impact on human health of our manipulation of bacterial flora); "rDNA Large-Scale Fermentation Experiments Should Have Flexible Physical Contain-

ment Requirements for 'Exempt' Organisms," *Blue Sheets,* February 22, 1984, pp. P, R–6 (describes the NIH's Recombinant DNA Advisory Committee's recommendation for flexibility in the commercial use of recombinant DNA technology—since filters employed commercially are only 99.97 percent effective in preventing release of genetically engineered organisms into the environment, the release of some genetically altered organisms, given current technology, is inevitable); T. Beardsley, "Bug Takes to Ski Slopes," *Nature,* 308 (1984), 303 (describes the potential use of bacteria capable of producing ice nucleation protein in order to improve snow-making on ski slopes—in this case, the bacteria will be killed since it is only the presence of the nucleation protein that is important for snow-making).

5. R. Thurow, "A Vine in Dixie Creeps Its Way into Infamy," *Wall Street Journal,* July 24, 1979, p. 1; J. L. Michael, "Some New Possibilities to Control Kudzu Pueraria-Lobata," *Proceedings of the Southern Weed Science Society,* 35 (1982), 237–40; R. M. May, "A Test of Ideas About Mutualism," *Nature,* 307 (1984), 410–11; P. A. Parsons (ed.), *The Evolutionary Biology of Colonizing Species* (Cambridge, 1983) (discusses the mode by which colonizing species evolve and survive, and holds interest for anyone trying to ascertain the end results of free-release experiments). See also the review by Nick Barton, "Genetics and Ecology," in *Nature,* 308 (1984), 87–88.

6. P. H. Abelson, "Chemistry and the World Food Supply," *Science,* 218 (1982), 333 (discusses the potential of cDNA technology to improve the world's food supply); Martin Alexander, "Ecological Consequences: Reducing the Uncertainties," *Issues in Science and Technology,* 1 (1985), 56–68.

7. "Genetic Engineering Raises Concerns of Senator Hatfield (R-Oregon)," 8.

8. R. E. Lenski, "Releasing Ice-Minus Bacteria," *Nature,* 307 (1984), 8. This paper discusses the use of specific bacteriophages to control the spread of engineered organisms. It is precisely this kind of control mechanism that should be available to society in the event that a released organism runs amuck.

Chapter X

1. P. Felig and M. Bergman, "Insulin Pump Treatment of Diabetes: Decision-Making Without Definitive Data," *JAMA,* 250 (1983), 1045–1047.

2. For examples of the difficulty involved in teasing out correct conclusions from large clinical studies, see Multiple Risk Factor Intervention Trial Research Group, "Multiple Risk Factor Intervention Trial: Risk Factor Changes and Mortality Results," *JAMA,* 248 (1982), 1465–77; P. Leren *et al.,* "MRFIT and the Oslow Study," *JAMA,* 249 (1983), 893–94; I. Holme *et al.,* "Treatment of Mild Hypertension with Diuretics: The Importance of ECG Abnormalities in the Oslow Study and in MRFIT," *JAMA,* 251 (1984), 1298–99. Multiple studies have demonstrated that the treatment of hypertension of the moderate or severe form can reduce complications and prolong life. The question of the appropriateness of treatment in the mild form is under investigation. Although the Hypertension Detection and Followup Study showed clear-cut benefit from the treatment of mild hypertension, at least one other, the so-called MRFIT (Multiple Risk Factor Intervention

Trial) study, found that mortality was actually increased in a subset of the mild hypertensives having abnormal electrocardiograms initially. The possible reasons for this latter finding are complex and suggest the need for additional trials employing different drugs. In the meantime, clinicians are forced to make judgments regarding treatment of mild hypertension based on the known studies, the known epidemiology of the process, and their understanding of the cellular biology of the disease.

Chapter XI

1. For a detailed discussion of the critical need of the domestic pharmaceutical industry for innovation and a discussion of the precarious position of the domestic industry vis-à-vis foreign competition, see L. H. Sarett, "Research and Invention," *PNAS*, 80 (1983), 4752–54.

2. W. S. Moore, "Theoretical and Physical Background to NMR Scanning," in *NMR Imaging: Proceedings of an International Symposium on Nuclear Magnetic Resonance Imaging, Held at the Bowman Gray School of Medicine of Wake Forest University, Winston-Salem, North Carolina, October 1–3, 1982* (Winston-Salem, N.C., 1984), 1–4.

3. M. R. Willcott, J. J. Ford, and G. E. Martin, "Nuclear Magnetic Resonance Spectroscopy," in *ibid.*, 5–14; I. L. Pykett *et al.*, "Principles of Nuclear Magnetic Resonance Imaging," *Radiology*, 143 (1982), 157; M. J. Pramik and C. Laino, "Already NMR Scans Bare Heart Vessels," *Medical Tribune*, 25 (January 25, 1984), 1 (describes the growing ability of NMR imagers to visualize the beating heart utilizing time-gating mechanisms). The potential for NMR to identify early lesions in the heart and vasculature noninvasively (in fact, even without contrast) has enormous clinical implications for the practice of cardiology, neurology, and vascular surgery.

4. K. L. Behar *et al.*, "High-Resolution ^{1}H Nuclear Magnetic Resonance Study of Cerebral Hypoxia *in Vivo*," *PNAS*, 80 (1983), 4945–48 (lays the groundwork for the possible study of brain chemistry and more particularly for the diagnosis of ischemia in brain utilizing nuclear magnetic resonance techniques); G. K. Radda *et al.*, "Clinical Applications of ^{31}P NMR," in *NMR Imaging*, 159–69; G. K. Radda *et al.*, "Examination of a Case of Suspected McArdle's Syndrome by ^{31}P Nuclear Magnetic Resonance," *NEJM*, 304 (1981), 1338–42; "Lilly Cancer Research Includes Work on Magnetic Microspheres," *Blue Sheets*, April 27, 1983, pp. P, R–9 (explains how an anticancer drug was formulated into tiny spheres containing iron oxide and a protein—this fledgling attempt may herald the dawn of a new age of electronic surgery).

5. In the past, medicine was known as *physic* and only later was the term *physics* employed in its modern sense. Medicine may recapture it yet. See Lewis Thomas, *The Lives of a Cell: Notes of a Biology Watcher* (New York, 1974), 154–55.

Chapter XII

1. P. F. Drucker, "Quality Education: The New Growth Area," *Wall Street Journal,* July 19, 1983, p. 28.

Chapter XIII

1. W. F. Doolittle and C. Sapienya, "Selfish Genes: The Phenotype Paradigm and Genomic Evolution," *Nature,* 284 (1980), 601–603; L. E. Orgel and F. H. C. Crick, "Selfish DNA: The Ultimate Parasite," *Nature,* 284 (1980), 604–607; W. R. Jelinek and C. W. Schmid, "Repetitive Sequences in Eukaryotic DNA and Their Expression," *Annual Review of Biochemistry,* 51 (1982), 813; W. R. Jelinek *et al.,* "Ubiquitous Interspersed Repeated DNA Sequences in Mammalian Genomes," *PNAS,* 77 (1980), 1398; J. J. Krolewski, "Members of the Alu Family of Interspersed Repetitive DNA Sequences Are in the Small Circular DNA Population of Monkey Cells Grown in Culture," *Journal of Molecular Biology,* 154 (1982), 399; C. W. Schmid and W. R. Jelinek, "The Structure and Organization of the Major Interspersed Repetitious Sequence in Mammalian DNA: The Alu Sequence," *Science,* 216 (1982), 1065; F. P. Rinehart, "Renaturation Rate Studies of a Single Family of Interspersed Repeated Sequences in Human DNA," *Biochemistry,* 20 (1981), 3003 (makes the point that there are about 500,000 members of the Alu family in the human genome, each apparently about 300 base pairs in length); Andrew Leigh Brown, "On the Origin of the Alu Family of Repeated Sequences," *Nature,* 312 (1984), 106 (describes how the Alu-sequences may have arisen from so-called 7SL RNA and then multiplied in the genome).

2. F. M. Stahl, "Special Sites in Generalized Recombination," *Annual Review of Genetics,* 13 (1979), 7; M. Collins and G. M. Rubin, "The Structure of Chromosomal Rearrangements Induced by the FB Transposable Element in Drosophila," *Nature,* 308 (1984), 323–27 (describes the foldback [FB] family of transposable elements—one repetitive DNA sequence that appears to have an important function in genetic rearrangement); J. R. S. Fincham, "Transposable Elements and Plant Gene Structure," *Nature,* 306 (1983), 425; M. Kress *et al.,* "Functional Insertion of an Alu Type 2 (B2 SINE) Repetitive Sequence in Murine Class I Genes," *Science,* 226 (1984), 974–77; J. G. Sutcliffe *et al.,* "Identifier Sequences Are Transcribed Specifically in Brain," *Nature,* 308 (1984), 237–41. Indeed, one repeated nucleotide sequence appears to be the identifier sequence for brain proteins. It is entirely possible that other repeated sequences serve roles still to be appreciated. The role, or lack thereof, of these "identifier sequences" remains problematic. See C. Sapienzia and B. St.-Jacques, "'Brain-Specific' Transcription and Evolution of the Identifier Sequences," *Nature,* 319 (1986), 418–20.

3. Stephen J. Gould, *Hens' Teeth and Horses' Toes: Further Reflections in Natural History* (New York, 1983), 166–76. Gould here takes up the question of the various forms of repetitive DNA and concludes that, in his view, the selfish DNA hypothesis best fits the facts. In speaking of the relationship between selfish

DNA and individual human beings, he states, "Nature, however, acknowledges many kinds of individuals, both great and small."

Chapter XIV

1. B. L. Strehler, *Time Cells and Aging* (2nd ed.; New York, 1977). This book is an excellent summary of experimental and theoretical work dealing with the nature of aging and its prevention.

2. *Ibid.*, 39, 167.

3. *Ibid.*, 37–46.

4. *Ibid.*, 214; J. L. Fox, "Bamboo Loss Endangers Giant Pandas in China," *Science*, 223 (1984), 466–67. Bamboo ages in parallel—that is, all plants that are members of the same species of bamboo are members of the same age cohort and are in the same portion of the aging cycle at any given time. Periodically all such bamboo dies out, not to be seen again until the seeds left behind as a cohort begin to grow. This total loss of the adult bamboo of a given species has definite ecological implications, particularly when alternative species of bamboo have been destroyed by human habitation. The giant panda utilizes bamboo as a major food source, and thus periodically pandas are placed in jeopardy by the synchronous aging of bamboo. The curiosity then becomes an environmentally important issue. See also J. R. Smith *et al.*, "Intraclonal Variation in Proliferative Potential of Human Diploid Fibroblasts: Stochastic Mechanism for Cellular Aging," *Science*, 203 (1980), 82–84; T. B. L. Kirkwood, "Cell Proliferation: Paths to Immortality and Back," *Nature*, 308 (1984), 226 (reviews recent research into mortality and immortality in cell culture, and discusses the interesting observation that immortal cell lines can be converted to mortal cell lines—apparently, immortality is not "forever."

5. T. Schreiner, "Aging, Low Births Hit Frost Belt," *USA Today*, October 10, 1984, pp. 1, 5 (notes that in the United States as a whole [not just the frost belt] the fertility rate is 1.8 children per woman, less than the 2.1 needed to sustain the population); C. M. Taleuber, "America in Transition: An Aging Society," in U.S. Census Bureau, *Current Population Reports* (Washington, D.C., 1983), Ser. E-12, No. 128; S. A. Schroeder *et al.*, "Frequency and Clinical Description of High Cost Patients in 17 Acute Care Hospitals," *NEJM*, 300 (1979), 1306–1309; J. F. Fries, "Aging, Natural Death and the Compression of Morbidity," *NEJM*, 310 (1984), 659–60; E. L. Schneider and J. A. Brody, "Aging, Natural Death and the Compression of Morbidity," *NEJM*, 310 (1984), 660. In this letter the authors comment that by the year 2050 the population aged eighty-five and over will have increased from 2.3 million to approximately 16 million. Currently, approximately 20 percent of persons over eighty-five have severe cognitive impairment and require long-term care. The authors use these data to argue strongly that great effort must be given over to reducing the morbidity of aging. Discussions regarding the effect of enhanced longevity on national economics can sometimes become quite heated, if not enlightening. See, for example, I. Peterson, "Colorado Governor Aims to Provoke," New York *Times*, April 1, 1984, p. 22. This article describes the remarks of

234

Governor Lamm, taken out of context to be sure, suggesting that the elderly terminally ill have a "duty to die" in order to save resources for the coming generation.

6. S. W. Mosher, "Why Are Baby Girls Being Killed in China?" *Wall Street Journal,* July 25, 1983, p. 11.

7. R. Bucalla *et al.,* "Modification of DNA by Reducing Sugars: A Possible Mechanism for Nucleic Acid Aging and Age-Related Dysfunction in Gene Expression," *PNAS,* 81 (1984), 105–109 (describes the effect of glucose on nucleic acid chemistry and one mechanism by which nucleic acid dysfunction occurs, a "browning reaction" in which glycosylation of nucleic acids produces molecular morbidity); B. N. Aims, "Dietary Carcinogens and Anticarcinogens," *Science,* 221 (1983), 1256–64 (discusses the idea that DNA damage resulting in cancer can be altered by diet; the reader is left to infer that similar dietary changes may play a role in any DNA mechanism involved in aging). "How People Will Live To Be 100 or More: An Interview with Dr. Roy Walford, Professor of Pathology at the University of California at Los Angeles Medical School," *U.S. News and World Report,* July 4, 1983, pp. 73–74; T. Lloyd, "Food Restriction Increases Lifespan of Hypertensive Animals," *Life Sciences,* 34 (1983), 401–407; O. M. Pereira-Smith and J. R. Smith, "Evidence for the Recessive Nature of Cellular Immortality," *Science,* 221 (1983), 964–66. An alternative interpretation of these data is that at least some forms of "immortality" are genetically dominant. In any case, a genetic mechanism is implicated in aging.

Chapter XV

1. C. E. Shannon and W. Weaver, *The Mathematical Theory of Communication* (Urbana, Ill., 1949); J. R. Pearce, *Symbols, Signals, and Noise* (New York, 1961).

2. In fact, the tenets of quantum mechanics have been explicitly incorporated into communications theory. Shannon determined that the maximum rate of information transmission (C) was related to average signal power (P), bandwidth (B) and noise-power (N) by the formula:

$$C = [\log_2(1 + \bar{P}/N)] \, B$$

It can, however, be shown that the uncertainty principle of quantum mechanics requires that a certain obligatory noise-power of quantum mechanical origin (QN) be associated with any attempt to send information. Moreover, QN can be calculated from:

$$QN \cong \hbar \omega B$$

where $\hbar$ is Planck's constant/2π and ω is frequency. Thus,

$$C \cong B \log_2 [1 + \bar{P}/(N' + QN)]$$

with QN representing quantum mechanical noise and N' classical noise. For a detailed analysis of the interface between quantum mechanics and information theory, see A. Yariv, *An Introduction to Theory and Applications of Quantum Mechanics* (New York, 1982), 32–35. Although our use of information theory as a guide to quantum reality is only an approximation, it is interesting to note that the in-depth analysis of "classical problems" can lead to virtually the same conclusions as those reached on the basis of standard quantum methods. See, for example, J. Maddox,

"Brownian and Quantum Motion," *Nature,* 311 (1984), 101. A current controversy stemming in part from the interrelationship between information, energy, and quantum mechanics centers on the question, Is there in principle a minimum energy requirement for the performance of a computation by a calculator? Or, to put this another way, If friction, resistance, heat loss, and noise are excluded, must a computer be supplied with energy to perform some or all computations? Some argue that a minimum energy of kT per bit of information processed is required, while others contend that this energy per computation can in principle be reduced to a level arbitrarily close to zero. The controversy itself raises interesting questions about the nature of computation. Is information after computation in any thermodynamic sense different from information prior to computation? Probably not always. Are the mechanics of computation "irreversible" in all cases? Probably not. See Arthur Robinson, "Computing Without Dissipating Energy," *Science,* 223 (1984), 1164–66. It should also be noted, however, that the rate of signal flow in a computer is limited to E/h bits/sec where E is the total energy available for signaling. See J. H. Bremermann, "Complexity and Transcomputability," in R. Duncan and M. Weston-Smith (eds.), the *Encyclopedia of Ignorance* (New York, 1977), 172. For practical as well as theoretical reasons there are thus quantum mechanical limits on the rate of computing.

3. For a more detailed discussion of the possible origins of the second law as well as its relationship to information theory and the intrinsic irreversibility of stochastic processes, see P. C. W. Davies, "Inflation in the Universe and Time Asymmetry," *Nature,* 312 (1984), 524–27; and H. Everett III, "Theory of the Universal Wave Function," in B. S. DeWitt and N. Graham (eds.), *The Many-Worlds Interpretation of Quantum Mechanics* (Princeton, 1973), 30–32. In addition, it should be noted that in a real sense there are two interpretations of the second law of thermodynamics, one derived phenomenologically from observations performed on macroscopic systems and the other derived from statistical mechanics and probability considerations. For a discussion of the two interpretations of the second law, the interested reader is referred to B. H. Lavenda, "Brownian Motion," *Scientific American,* 252 (1985), 70–85.

4. C. J. A. Game, "BVP Models of Nerve Membrane," *Nature,* 299 (1982), 375 (this and following letters demonstrate the tremendous mathematical sophistication being applied to the modeling of nerve conduction, another facet of understanding the nervous system); N. V. Swindale, "Anatomical Logic of Retinal Nerve Cells," *Nature,* 303 (1983), 570–71. For an interesting discussion of probability, individuality, and collective thought, see Lewis Thomas, *The Lives of a Cell: Notes of a Biology Watcher* (New York, 1974), 165–69. The author makes an interesting point that virtually none of us thinks as an isolated entity but rather as one element of a network involving potentially millions of other individuals. He further asserts that the thought process is still evolving and is actually at a "prokaryotic" stage. He seems to long for the advent of "eukaryotic" thought. See also M. Mishina *et al.,* "Expression of Functional Acetylcholine Receptor from Cloned DNAs," *Nature,* 307 (1984), 604–608 (demonstrates how recombinant DNA technology can lead to

a better understanding of the function of neural transmitter receptors and therefore potentially of nervous system functioning); J. L. Fox, "Gene Splicers Contemplate Rat Brain: Genes Active in the Brain May Carry a Tag in Noncoding Regions and a Potential New Transmitter Has Been Identified," *Science,* 222 (1983), 828–29. The interrelationship between thought, consciousness, and genetics can be found on many levels; see, for example, G. Kolata, "Math Genius May Have Hormonal Basis," *Science,* 222 (1983), 1312.

5. J. R. Sanes, "More Nerve Growth Factors," *Nature,* 307 (1984), 500. This report discusses the multiplicity of growth factors known to affect the growth and development of nerve cells and suggests additionally the possibility that one particular growth factor may be inhibited by autoantibodies in patients with amyotrophic lateral sclerosis—so-called Lou Gehrig's disease. If confirmed, this would be an exciting cellular observation that may be extrapolated to clinical medicine. Additionally, the existence of nervous system growth factors greatly extends medicine's capacity to explain the development of the nervous system and perhaps the genesis of thought. See also "Brain Healing: Implanting Fetal Cells in Rats," *Time,* August 8, 1983, p. 59; G. Kolata, "Brain Grafting Work Shows Promise," *Science,* 221 (1983), 1277; F. H. Gage *et al.,* "Aged Rats: Recovery of Motor Impairments by Intrastriatal Nigral Grafts," *Science,* 221 (1983), 966–69 (describes what amounts to brain cell transplants for the therapy of disorders of brain function; its implications are very important).

6. G. A. Miller, *Psychology: The Science of Mental Life* (New York, 1962). See pp. 163–64 for a discussion of the engram and a description of an experiment of Karl Lashly, which suggests that each engram is widely distributed in the brain. Recent work raises the possibility that engrams are stored in the brain much as holograms are recorded on film. That is, the engram may be a pattern of neuronal activity, any part of which can approximate an entire memory. This model could prove to be an important clue in understanding the mechanisms of memory.

Chapter XVI

1. C. Darwin, *The Origin of Species and the Descent of Man* (New York, 1936); D. C. Queller, "Sexual Selection in Hermaphroditic Plants," *Nature,* 306 (1983), 706–707; E. Mayr, *The Growth of Biological Thought* (Boston, 1983) (an important book expressing the views of one of the founders of the modern synthesis theory of Darwinian evolution).

2. J. Rifkin and N. Perlas, *Algeny* (New York, 1983), 155, 125; Stephen J. Gould, *Hens' Teeth and Horses' Toes: Further Reflections in Natural History* (New York, 1983). Gould's superb book describes the successes, failures, and uncertainties in present-day evolution theory. In an interesting essay, "Evolution as Fact and Theory" (pp. 253–62), the author makes the point that criticizing the modern synthesis of Darwinian theory and suggesting (correctly or incorrectly) the theory of punctuated equilibria in no way implies that the basic concept of evolution is incorrect. These additional theories should be viewed as refinements of Darwin's ideas, not true replacements.

3. Gould, *Hens' Teeth*, 335. The author here discusses the methods for developing the amount of genetic variation in natural populations and concludes that "most populations maintain too much variation to support the usual claim that all genes are scrutinized by natural selection." This leads to the idea of *neutral* genes. On p. 337, Gould discusses ways by which new species can form by so-called "chromosomal speciation." In this view, some social organizations (genetically determined?) facilitate the origin of new species by "rapid and accidental chromosomal change." The existence among kin groups of harems in which a dominant male mates with multiple females with the possibility of brother-sister mating among the offspring could protect new species produced by this mechanism. See also D. Penny, "How 'Gradual'?" *Nature,* 307 (1984), 8–9 (defends traditional Darwinism against the attacks of those espousing punctuated equilibria); J. B. S. Haldane, *The Causes of Evolution* (Ithaca, N.Y., 1966), 3, 110; Rifkin and Perlas, *Algeny,* 135, 151–53.

4. Haldane, *Causes of Evolution,* 142–43.

5. P. Leder, "Mechanisms of Gene Evolution," *JAMA,* 248 (1982), 1582–91 (describes what is currently known about genetic mechanisms, including jumping genes, and relates this information to evolution); D. Reanney, "Genetic Noise in Evolution?" *Nature,* 307 (1984), 318–19. The latter paper discusses error-prone communication between DNA and RNA, as well as the reverse communication between RNA and DNA. From an analysis of these processes, it can be concluded that yet another factor in evolution is error-prone synthesis of DNA based on cellular RNA. The evolutionary process therefore can be seen to be very complex.

6. G. Dover, "Molecular Drive: A Cohesive Mode of Species Evolution," *Nature,* 299 (1982), 111–17. This interesting review describes the possibility that intra- and interchromosomal DNA shifts produce "a new form of evolution for natural selection to work on." The author points out that "in many families of genes and noncoding sequences, however, fixation of mutations within a population may proceed as a consequence of molecular mechanisms of turnover within the genome. These mechanisms can be both random and directional in activity. There are circumstances in which unusual concerted pattern of fixation permits the establishment of biological novelty and species discontinuity in a manner not predicted by the classical genetics of natural selection and genetic drift." The reader interested in molecular drive is strongly advised to read this paper. Of course, any theory of evolution and its rate of progression must consider cataclysmic environmental events. See, for example, W. Alvarez *et al.,* "Impact Theory of Mass Extinction and the Invertebrate Fossil Record," *Science,* 223 (1984), 1135–41. This paper summarizes the evidence supporting the idea that the impact of large meteors upon the earth on a regular basis as a result of the sun's rotation around the center of the galaxy could account for the mass extinction of life forms such as the dinosaurs.

7. R. Lewin, "Some Avian Puzzles Solved," *Science,* 226 (1984), 1180. This interesting article describes how modern DNA hybridization techniques have overturned long-held but erroneous ideas about the origins of various species of birds. Lawrence J. Henderson, in *The Fitness of the Environment: An Inquiry into the*

Biological Significance of the Properties of Matter (Boston, 1958), 288, states that "from the earliest days of the new hypothesis, it has been widely recognized that to accept the survival of the fittest as one factor in the adaptation of life to its environment is quite a different matter from proving it to be the only force which directs evolution." See also Stephen J. Gould, *The Panda's Thumb: More Reflections in Natural History* (New York, 1980), 179–85, for a discussion of the Eldredge-Gould model of "punctuated equilibria"; and L. L. Cavalli-Sfozra and M. W. Feldman, "Cultural Versus Genetic Adaptation," *PNAS*, 80 (1983), 4993–96, which makes the point that evolution is intrinsically more complex than even biology would suggest, and investigates the impact of cultural as well as genetic adaptation on selection.

Chapter XVII

1. "Fertilization, Embryo Transfer Procedures Raise Many Questions," *JAMA*, 252 (1984), 877–82.

2. R. W. Sperry, "Problems Outstanding in the Evolution of Brain Function," in R. Duncan and M. Weston-Smith (eds.), *The Encyclopedia of Ignorance* (New York, 1977), 423–33; T. E. Adrian *et al.*, "Neuropeptide Y Distribution in Human Brain," *Nature*, 306 (1983), 584–86; D. T. Kreiger, "Brain Peptides: What, Where, and Why?" *Science*, 222 (1983), 975–85; N. V. Swindale, "Anatomic Logic of Retinal Nerve Cells," *Nature*, 303 (1983), 570–71. These articles illustrate the tremendous potential complexity of the brain as is being elucidated by the use of specific peptide and gene probes.

3. R. H. Scheller and R. Axel, "How Genes Control an Innate Behavior," *Scientific American*, 250 (1984), 54–62; N. Chaudhari and W. E. Hahn, "Genetic Expression in the Developing Brain," *Science*, 220 (1983), 924–28; M. Mishina *et al.*, "Expression of Functional Acetylcholine Receptor from Cloned DNAs," *Nature*, 307 (1984), 604–608. These reports of molecular research into the area of brain function touch on the two major areas of recent progress—recombinant DNA technology applied to the brain, and immunological study of neurotransmitters. Of particular note is the discovery that the mechanism for the transcription of genes related to neural functioning seems to be different from that found elsewhere in the body. More particularly, in the brain genes appear to be initially partially transcribed by RNA polymerase-III and only subsequently is the entire gene transcribed by message-related RNA polymerase-II. The significance of this mechanism is at present unknown.

4. D. Hofstadter, *Gödel, Escher, Bach: An Eternal Golden Braid* (New York, 1980), 594–603, 709–715; M. M. Waldrop, "The Necessity of Knowledge: The Essence of Intelligence Seems To Be Less a Matter of Reasoning Ability Than of Knowing a Lot About the World," *Science*, 223 (1984), 1279–82 (focuses on the question of whether reasoning ability is in fact determined by symbolic logic networks or whether, on the other hand, it is a more eclectic entity derived in large part from experience); R. L. Gregory, "Consciousness," in Duncan and Weston-Smith (eds.), *Encyclopedia of Ignorance*, 273–81 (addresses the issue of how one really

knows if a machine is conscious); M. M Lehman, "Human Thought and Action as an Ingredient of System Behavior," in *ibid.*, 347–54.

5. E. P. Wigner, *Symmetries and Reflections: Scientific Essays* (Cambridge, Mass., 1970). In these essays the author discusses, among other things, the probability of the existence of a self-reproducing unit as viewed from the vantage point of quantum mechanics. Wigner takes the view that the interaction of mind and matter is quantum mechanically distinct from matter/matter interactions (pp. 179–80). Wigner also discusses Elsasser's view of so-called biotonic laws as they relate to the interaction of consciousness and matter. See W. M. Elsasser, *The Physical Foundations of Biology* (London, 1958). The basic Elsasser argument is that living cells are in principle too complex to permit their behavior to be calculated even on the basis of quantum mechanics. In support of this view he notes that a computer that could store all the information contained in the germ cells of man would be inconceivably large. See also A. Cottrell, "Emergent Properties of Complex Systems," in Duncan and Weston-Smith (eds.), *Encyclopedia of Ignorance*, 129–35. This essay discusses quantum effects in biological processes and the complexity of living systems. For a somewhat different approach to the problem of knowability, see T. L. Clarke, "Limitations of Physical Theory," *Nature*, 308 (1984), 20. Here it is argued that because the traditional interpretation of quantum mechanics requires a collapsed state at the time of observation—that is, observation is an intrinsic part of prediction—quantum mechanics per se can never completely explain conscious observation. This is essentially a variation on Gödel's theorem, and the author goes on to demonstrate just how "Gödelian" our world is. The author similarly touches upon the anthropomorphic interpretation of Dirac's large numbers. That is, the argument is discussed that the constants of nature have the values they do because if they didn't, no life could have evolved to contemplate them. The author concludes, "Any comprehensive physical theory is necessarily incomplete because of Gödel's theorem. . . . Determining the actual outcome requires another level of observation outside the physics in analogy to the resolution of incompleteness in mathematics by adding new axioms." It should be pointed out, however, that although this latter conclusion seems incontrovertible, the author's use of the observation problem as a strict limit on reductionism does not apply in the case of the "many-worlds" interpretation of quantum mechanics (see chapter 20). For an analysis of the possible relationship of physics and Eastern philosophy, see Fritjof Capra, *The Tao of Physics* (New York, 1984).

Chapter XVIII

1. L. H. Sarett, "Research and Invention," *PNAS*, 80 (1983), 4752–72.

2. Recent evidence indicates that cyclosporin A inhibits the synthesis of a T-cell growth factor called interleukin-2 and that this could account for at least part of its immunosuppressive action. How it accomplishes this inhibition of synthesis on a molecular basis remains unknown. See J. F. Elliott *et al.*, "Induction of Interleukin 2 Messenger RNA Inhibited by Cyclosporin A," *Science*, 226 (1984), 1439–41. Another drug whose development makes this same point (*i.e.*, that useful phar-

maceuticals may first appear as molecular biological curiosities) is suramin. This drug, useful in the treatment of a parasitic disease, was found to have the capacity to block the reverse transcriptase of many retroviruses. Because the significance of retroviruses and reverse transcriptase for human disease was not appreciated at the time of this discovery, years passed before the drug was eventually tried against the killer disease AIDS. See H. Mitsuya *et al.*, "Suramin Protection of T-Cells *in Vitro* Against Infectivity and Cytopathic Effect of HTLV-III," *Science,* 226 (1984), 172–74. For the story of cyclosporin, see G. Kolata, "Drug Transforms Transplant Medicine," *Science,* 221 (1983), 40–42.

3. J. M. Hudak *et al.*, "Discovery and Development of Flecainide," *AJC,* 53 (1984), 17B–20B. This article describes the discovery and development of a typical drug by the pharmaceutical industry. As the authors state: "Drugs have primarily two origins. Naturally occurring drugs are derived mainly from plants, bacteria, or other modified biological systems; synthetic chemicals are produced by medicinal chemists primarily in the pharmaceutical industry." What the authors fail to note is that synthetic chemicals produced by the pharmaceutical industries are developed either as analogs of natural compounds or as compounds with biological activity in some bioassay. This description of the discovery and development of a new drug is classical, and after studying it the reader may well agree that drugs will soon have a third origin—namely, one based on molecular biological effect with bioassay activity and clinical efficacy demonstrated only subsequently. Note particularly the paragraph on page 18B, which begins, "A Riker pharmacologist, Jack Schmid, astutely pointed out that since the two properties often go hand in hand, compounds with local anesthetic property should also be tested for antiarrhythmic activity. Screening of these compounds for antiarrhythmic activity began in 1968. The primary test used for determining antiarrhythmic activity was the mouse protection screen." The pattern of pharmaceutical development is well established and at present is no different for the development of an antiarrhythmic or of an immunosuppressant. The future will hold something quite different. K. Bouton, "Academic Research And Big Business: A Delicate Balance," *New York Times Magazine,* September 11, 1983, pp. 62–153; L. Derman, "Genetics: The Youngest Basic Science," *Harvard Medical Alumni Bulletin,* 57 (1983), 16–25; B. J. Culliton, "The Hoechst Department at Massachusetts General," *Science,* 216 (1982), 1200–1203.

Chapter XIX

1. Committee on Ethics of the American Heart Association, "Ethics of Biomedical Technology Transfer," *Circulation,* 67 (1983), 942A–46A. This is an excellent discussion of the ethics of medical technology, including recombinant DNA research and application. The authors conclude that, "having accepted the technology of recombinant DNA molecular research, society need not condone or even tolerate all possibilities of genetic manipulation, but choose among those and advance only those found acceptable. We have, however, arrived at a paradox of modern biomedical science: the technology which can and must be directed and ad-

vanced (but can never be reversed) by human choice threatens to redefine and redesign what humanity is and ought to be, thereby undermining the very concept of people as unique entities capable of shaping by individual choice, their own destinies." This is a view echoed by many church leaders and should challenge everyone to clarify the philosophical conception of the individual human being. Yet because of the Gödelian nature of the "individual," the threat from molecular biology may be more apparent than real.

2. Martin Heidegger, *Discourse on Thinking*, trans. J. M. Anderson and E. H. Freund (New York, 1966), 52–53.

3. D. Hofstadter, *Gödel, Escher, Bach: An Eternal Golden Braid* (New York, 1980).

4. H. W. Smith, *From Fish to Philosopher: The Story of Our Internal Environment* (New York, 1959). This classic work dealing with the fitness of our internal environment and the concept of physiological homeostasis contains in addition an intriguing chapter on the nature of consciousness and the cellular complexity that underlies it as well as all of life. See also J. Maddox, "New Twist for Anthropomorphic Principle," *Nature*, 307 (1984), 409. This interesting article reviews the position that things in many ways are as they are because the fact that we are here to observe them places large constraints on the physical reality about us. This is reminiscent of Lawrence J. Henderson's arguments in his classic book *The Fitness of the Environment: An Inquiry into the Biological Significance of the Properties of Matter* (Boston, 1958) and points out the intrinsic interrelation of the physical universe and man.

Chapter XX

1. Science has for some time been puzzled by the coexistence of the arrow of time, the second law of thermodynamics, biological evolution, and the formation of galaxies. In order to try to put these and other phenomena into the same theoretical framework, P. C. W. Davies argues that the inflation of the universe after the Big Bang led to certain conditions that could explain the time asymmetry we now see about us ("Inflation and Time Asymmetry in the Universe," *Nature*, 301 [1983], 398–400). D. N. Page, however, disagrees ("Inflation Does Not Explain Time Asymmetry," *Nature*, 304 [1983], 39–41). The matter is still open, to say the least. Exactly how evolution and mortality are implied by the second law is similarly open to some discussion. See also P. C. W. Davies, "Inflation in the Universe and Time Asymmetry," *Nature*, 312 (1984), 524–27; D. Park, *The Image of Eternity: Roots of Time in the Physical World* (Amherst, Mass., 1980). Is mortality decreed by the second law? See the review of Park's book by G. Uhlenbeck in *Nature*, 288 (1980), 305–306. Is evolution at odds with the second law of thermodynamics? Lest the reader be deluded into believing that everyone has accepted the power of thermodynamics, quantum mechanics, and the second law of thermodynamics, read the curious report in *Science*, 223 (1984), 571–72, in which Louisiana's Joseph Newman claims to have invented what amounts to a perpetual motion machine. New Orleans scientists feel it worth the time and effort to investigate the

claim. See also A. G. Leggett, "The Arrow of Time and Quantum Mechanics," in R. Duncan and M. Weston-Smith (eds.), *The Encyclopedia of Ignorance* (New York, 1977), 101–109; S. A. Bludman, "Thermodynamics and the End of a Closed Universe," *Nature*, 308 (1984), 319–22 (describes the use of thermodynamics and relativity physics to predict a plausible scenario for the future of the universe).

2. E. P. Wigner, *Symmetries and Reflections: Scientific Essays* (Cambridge, Mass., 1970) (discusses in detail the observational problem in quantum mechanics—the Stern-Gerlach experiment and the like); B. S. DeWitt, "Quantum Mechanics and Reality," *Physics Today*, 23 (1970), 155–65; J. Maddox, "Observation and Things Observed," *Nature*, 308 (1984), 601.

One of the more interesting approaches to the observation problem is that provided by the theory of quantum potentials, a theory that is mathematically consistent with traditional quantum mechanics. In this view, all masses are associated with a quantum field, which might be thought of as a track along which the particle might travel. Tracks of quantum potential corresponding to mutually exclusive outcomes (*e.g.*, one running up and one down) can coexist. The particle itself, however, actually travels on the tracks based on the amplitude of the quantum potential (the height of the rails) and random chance. Physical reality is associated only with the path the particle actually follows. The untraveled tracks serve only a bookkeeping role to help determine the future motion of the particle (that is, what new tracks the particle will encounter). According to this view, an observation made by my friend is made definitively, but the track the observed particle did *not* follow still produces uncertainty in *my* knowledge of what the observer saw, and I must take into consideration these untraveled tracks when I attempt to predict what I will observe. See D. J. Bohm, C. Dewdney, and B. H. Hiley, "A Quantum Potential Approach to the Wheeler Delayed-Choice Experiment," *Nature*, 315 (1985), 294–97.

3. R. M. Pirsig, "An Author and Father Looks Ahead at the Past," *New York Times Book Review*, March 4, 1984, pp. 7–8. In this piece, the author of *Zen and the Art of Motorcycle Maintenance* discusses the culture and philosophy of time and meaning.

4. H. Everett III, "Relative State Formulation of Quantum Mechanics," *RMP*, 29 (1957), 454–62; B. S. DeWitt and N. Graham (eds.), *The Many-Worlds Interpretation of Quantum Mechanics* (Princeton, 1973). This book is a compendium of the pertinent theoretical literature dealing with the "many-worlds interpretation" of quantum mechanics. It includes "Theory of the Universal Wave Function" by H. Everett III, the originator of the many-worlds view; and B. DeWitt, "Quantum Mechanics and Reality: Could the Solution to the Dilemma of Indeterminism Be a Universe in Which All Possible Outcomes of an Experiment Actually Occur?" (pp. 155–65). The latter essay addresses not only the famous thought experiment termed "Schrödinger's Cat," but also a discussion of so-called maverick worlds. DeWitt states, "All the worlds are there, even those in which everything goes wrong and all the statistical laws break down . . . if the initial conditions were right, the universe-as-we-see-it could be a place in which heat sometimes flows from cold bodies to hot. We can perhaps argue that in those branches in which the universe makes a

habit of misbehaving in this way, life fails to evolve; so no intelligent automata are around to be amazed by it." DeWitt argues that such maverick worlds must represent only a small percentage of all universes. However, if the totality of such universes is infinite, or if the vast majority of universes "run down," then should not the maverick worlds be viewed as essentially "everything—not only to those living in them but intellectually to all of us?" See also J. A. Wheeler, "Assessment of Everett's Relative State Formation of Quantum Theory," *RMP*, 29 (1957), 463–65.

5. A. Cottrell, "Emergent Properties of Complex Systems," in Duncan and Weston-Smith (eds.), *Encyclopedia of Ignorance*, 129–35. This essay discusses quantum effects in biological processes and the complexity of living systems.

Chapter XXI

1. J. Naisbitt, *Megatrends: Ten New Directions Transforming Our Lives* (New York, 1982), 27. Naisbitt writes: "During the first stage of technological innovation, technology takes the path of least resistance, that is, it is applied in ways that do not threaten people—reducing the chance that the technology will be abruptly rejected. The way society handled the introduction of microprocessors is a classic example of this first stage. The first application of the microprocessor was in toys. Who could object? Robots were first used in jobs considered unsafe or too dirty for humans. Who could object? Robots (for dangerous tasks) and toys represent the unthreatening path of least resistance. At the same time, this path has created a whole generation of computer-comfortable kids. In five years, young people entering the labor force will have had some form of information device in their hands most of their lives—from calculators to computer games to push button telephones." A similar phenomenon may be occurring in the area of molecular biology.

2. J. Rifkin and N. Perlas, *Algeny* (New York, 1983), 242–44. The question of the objectivity of science can be phrased in a different way. One can also ask, as J. M. Goldschvartz has, "Just how objective are scientists?" ("Lessons in Objectivity," *Nature*, 308 [1984], 8). See also the article to which Goldschvartz was responding, N. S. Hetherington, "Just How Objective Is Science?" *Nature*, 306 (1983), 727.

3. Reinhold Niebuhr, *The Children of Light and the Children of Darkness* (New York, 1944), 50–51.

4. It must always be remembered that, although reason is man's best friend, a certain Gödelian humility should infect us all. As Paul Doty said, "Because experts are burdened with too much knowledge, they have done poorly at predicting the future in science" (J. L. Fox, "The DNA Double Helix Turns 30," *Science*, 222 [1983], 30). For a discussion of some of the implications of, and evidence for, a "changing" gravitational constant G, see P. A. M. Dirac (ed.), *Directions in Physics* (New York, 1978), 71–92; and R. A. Matzner, "Summary on the Workshop on General Relativity in Astrophysics," *Annals of the New York Academy of Sciences*, 375 (1981), 449–58.

5. Naisbitt, *Megatrends*, 167.

6. Cotton Mather, *An Account of the Method and Success of Inoculating the Small-pox in Boston in New-England* (London, 1722), 8.

7. Martin Heidegger, *Discourse on Thinking,* trans. J. M. Anderson and E. H. Freund (New York, 1966), 52.

8. Zabdiel Boylston, *An Historical Account of the Smallpox Inoculated in New England, upon All Sorts of Persons, Whites, Blacks, and of All Ages and Constitutions* (2nd ed., London, 1730), 39.

Appendix

1. A. White, P. Handler, and E. Smith, *Principles of Biochemistry* (3rd ed.; New York, 1964), 204–218 (a good discussion of the theory of enzyme action); S. A. Bludman, "Thermodynamics and the End of a Closed Universe," *Nature,* 308 (1984), 319–22; A. G. Wilkinson *et al.,* "A Large Increase in Enzyme-Substrate Affinity by Protein Engineering," *Nature,* 307 (1984), 187–88 (describes efforts by molecular biologists to design more active enzymes—another novel application of Bioburst); T. H. Maugh II, "Catalysts That Break Nature's Monopoly: Chiral Complexes Can Approach the Specificity of Enzymes for Synthesis of Optically Active Compounds and Can Act on a Wide Variety of Substrates," *Science,* 221 (1983), 351–54 (discusses initial attempts at making synthetic enzymes).

2. D. Park, *The Image of Eternity: Roots of Time in the Physical World* (Amherst, Mass., 1980); P. C. W. Davies, "Inflation in the Universe and Time Asymmetry," *Nature,* 312 (1984), 524–27.

3. A. Lehninger, *Bioenergetics* (New York, 1965).

4. A. G. Leggett, "The Arrow of Time and Quantum Mechanics," in R. Duncan and M. Weston-Smith (eds.), *The Encyclopedia of Ignorance* (New York, 1977), 101–109.

5. For a discussion of how the genesis of life on earth might be reconciled with the second law of thermodynamics through the mediation of nonlinear kinetic laws of chemistry, see R. J. Field, "Chemical Organization in Time and Space," *American Scientist,* 73 (1985), 142–50. There are a few scientists who contend that life need not have evolved entirely (or at all) on earth, but rather that seeds of life may have traveled through space eventually to alight on the friendly earth—among other places. The origin of this life is unclear from the theory as presented. See, for example, D. A. J. Tyrrell, "A New Dimension to Evolutionary Theory?" *Nature,* 294 (1981), 489–90; Davies, "Inflation in the Universe and Time Asymmetry," 524–27. Arguments related to cosmology will likely undergo considerable refinement and/or revision as modern astronomy continues to develop a wealth of new data. For example, recent observations suggest that galaxies formed on the surfaces of expanding explosion-produced bubbles rather than as the result of gravitational contraction alone. This finding could have important implications for cosmology and all hypotheses related to it. See Walter Sullivan, "New View of Universe Shows Sea of Bubbles to Which Stars Cling," New York *Times,* January 5, 1986, p. 1.

Index

247

Index

Index

Digitalis, 48
Doolittle, Russell, 92
Douglas, William, 4
Dover, Gabriel, 171
Down's syndrome, 103
Dreyer, W. J., 43
Drucker, Peter, 144
Drug addicts, 98
Drugs, 48

Ecology, 191
Eco RI, 60
Eddington, Sir Arthur, 211
Education, 126
Egg, 161–63, 205, 206
Einstein, Albert, 194
Eldredge, Niles, 168
Electrons, 195
Electrophoresis, 70, 73
Elsasser, Walter, 179
Embryo, 76, 105, 107, 111, 112
Endonuclease restriction, 60, 62, 70–73
Endothelial cells, 81, 83
Energy, 23, 157, 162, 165, 205, 208, 209; kinetic, 206; potential, 205, 206; free, 157, 207, 209; activation, 22, 23, 207–209; electromagnetic, 136
England, xiv, 1, 168
Engrams, 164, 165
Enkephalin, 175
Enthalpy, 205
Entropy, 159, 194, 206–208, 210, 211
Environment, 190–92
Environmental Protection Agency, 121
Enzymes, 17, 19, 20, 22, 25, 28, 29, 30, 60–62, 93, 205, 208
Equilibrium, 193, 206–207
Escherichia coli (E. coli.), 62, 115, 116, 117, 124, 151
Ethiopia, 201
Eugenics, 103, 112, 113
Eukaryote, 36
Everett, Hugh, III, 196, 197
Evolution, 93, 106, 146, 153, 167–71

Evolutionary bottleneck, 105, 154
Exon, 74

Facilitators, 29, 35
Fetus, 43
Fibroblasts (3T3 cells), 89
Force: grand unifying, 210; electromagnetic, 211; strong nuclear, 211; weak nuclear, 211
Fossil record, 167
France, 99, 100
Franklin, Benjamin, 5
Franklin, James, 5
Friedman-Kien, A. E., 97
Future shock, 188

Galaxies, 208
Galen, 6
Gallo, Robert, 95, 100
Garrod, Archibald, 25
Gazette, 5
Geiger counter, 196, 197
Gene, 30, 43, 45
Gene sequencing, 68–72
Genetic change, dose of, 105, 113, 191, 201
Genetic code, 26, 27
Genetic drift, 170
Genetic engineering, 67, 93, 103, 106, 108, 112
Genome, 42, 44, 145, 191
Germ line, 103, 104, 145, 152, 153, 154, 201
Gestalt, 52
Gilbert, Walter, 68
Glucose, 19, 22, 23, 84, 208
Gödel, Escher, Bach, 177, 189
Gödel, Kurt, 8, 178, 179, 189, 190
Gonadtropin, 145
Gonococcus, 57
Gould, Stephen Jay, 147, 168
Gravitational constant, G, 203
Gravity, 211
Growth factors, 82–85, 86, 87, 92, 131, 163
Guanine, 15, 17
Gyrase, DNA, 183

249

Index

251

Index

252

Index

Index